S0-AKX-891

THEORY OF SOLITONS IN INHOMOGENEOUS MEDIA

WILEY SERIES IN NONLINEAR SCIENCE

Series Editors: **ALI H. NAYFEH, Virginia Tech**
ARUN V. HOLDEN, University of Leeds

Abdullaev	THEORY OF SOLITONS IN INHOMOGENEOUS MEDIA
Nayfeh	METHOD OF NORMAL FORMS
Nayfeh and Balachandran	NONLINEAR DYNAMICS: CONCEPTS AND APPLICATIONS
Nayfeh and Pai	LINEAR AND NONLINEAR STRUCTURAL MECHANICS
Rozhdestvensky	MATCHED ASYMPTOTICS OF LIFTING FLOWS

THEORY OF SOLITONS IN INHOMOGENEOUS MEDIA

Fatkhulla Abdullaev

Physical-Technical Institute,
Tashkent, Uzbekistan

JOHN WILEY & SONS
Chichester · New York · Brisbane · Toronto · Singapore

Baffins Lane, Chichester,
West Sussex PO19 1UD, England
Telephone: National Chichester (0243) 779777
International +44 243 779777

Other Wiley Editorial Offices

John Wiley & Sons, Inc., 605 Third Avenue,
New York, NY 10158-0012, USA

Jacaranda Wiley Ltd, 33 Park Road, Milton,
Queensland 4064, Australia

John Wiley & Sons (Canada) Ltd, 22 Worcester Road,
Rexdale, Ontario M9W 1L1, Canada

John Wiley & Sons (SEA) Pte Ltd, 37 Jalan Pemimpin #05-04,
Block B, Union Industrial Building, Singapore 2057

Library of Congress Cataloging-in-Publication Data

Abdullaev, F. Kh. (Fatkhulla Khabibullaevich)
Theory of solitons in inhomogeneous media / Fatkhulla Abdullaev.
p. cm.—(Wiley series in nonlinear science)
Includes bibliographical references and index.
ISBN 0 471 94299 5
1. Solitons. 2. Nonlinear theories. I. Title. II. Series.
QC174.26.W28A24 1994
530.1'24—dc20 93-29288
CIP

British Library Cataloguing in Publication Data

A catalogue record for this book is available from the British Library

ISBN 0 471 94299 5

Typeset in 11/13pt Times by Thomson Press (India) Limited, New Delhi
Printed and bound in Great Britain by Biddles Ltd, Guildford, Surrey

CONTENTS

INTRODUCTION

In recent years a new field of theoretical physics—soliton theory—is being rapidly developed. Great success has been attained in th elaboration of novel approaches to soliton theory. In particular, a classical inverse scattering transform (IST) has been developed, several problems have been solved on the basis of the quantum IST, complete integrability of some multidimensional nonlinear wave equations has been proved, etc.

The construction of the foundation for soliton theory stimulated numerous applications to the solution of nonlinear problems in condensed matter theory, theory of elementary particles, plasma physics, hydrodynamics, and other fields of physics.

The application of soliton theory to solid state problems in modern statistical physics was motivated by the need to take into account strong nonlinearity in the wave process investigated. In the case of classical nonlinear wave systems we encounter such a situation when nonlinear terms in wave equations affect a wave in the same manner as dispersive terms do. In the case of quantum systems, taking nonlinearity into account gets more important when the energy of interaction between elementary excitations of the system (quasiparticles) appears to be of the same order as the intrinsic energy of quasiparticles. Clearly, under such conditions conventional approximate methods, based on the assumption that the effect of system nonlinearity consists in the renormalization of parameters of the initial quasilinear modes, their final lifetime, and so on, are invalid for describing nonlinear systems.

Thus, in strongly nonlinear systems of solid state physics it is often very useful to proceed with soliton representations. In describing strongly nonlinear systems, solitons start to play the same role as quasi-particles do in describing weakly nonlinear systems.

At present the main results for the description of nonlinear processes in solids based on soliton theory were obtained with the application of idealized models to be solved by the IST (of the type of the nonlinear Schrödinger equation (NLS), sine–Gordon, Toda chain, etc.). There are many reviews and monographs [I.1–I.6], in which the general soliton theory is stated and the application of soliton concepts to solid state physics is described. Thus, in their monograph *Nonlinear Magnetization Waves. Dynamic and Topological Solitons* [I.1] Kosevich *et al.* analyzed the dynamics of magnetic soliton systems in ideal one- and multidimensional magnetic media. In his monograph *Solitons in Molecular Systems* [I.2], Davydov described solitons in molecular chains and their applications to biophysical problems. This monograph actually already contains some results on the dynamics of solitons under regular perturbations. Dynamics of optical solitons has been considered recently in books [I.9], [I.10]. Mathematical aspects of soliton theory are reported in the books by Zakharov *et al.* [I.3] and Ablowitz and Segur [I.4]. In this connection, the investigation of the physics of solitons in *real* systems is a topical problem. In this case, the soliton dynamics are described by equations frequently reduced to different kinds of perturbations of the initial completely integrable nonlinear wave equations. Such investigations include the problems of taking into account the interactions between different physical fields and solitons, the influence of defects, impurities, and structure imperfections on soliton dynamics, and the effect of a thermal bath on solitons. Also of interest is the study of interaction processes, and soliton propagtion in various inhomogeneous media such as transient radiation and soliton scattering. Interest in these problems is largely due to the investigation of mechanisms of energy and charge transfer in conducting polymers (of the polyacetylene type (CH_X)) and biopolymers, dynamics of domain walls in magnetics and systems with structural phase transitions, theory of optical solitons in inhomogeneous and active media, fluxons in inhomogeneous long Josephson junctions, etc.

To solve these problems it is necessary to investigate dynamics of solitons and nonlinear periodic waves in inhomogeneous condensed media. Results obtained along these lines are outlined in the present monograph. Note that in recent review Kivshar and Malomed [I.7] and book Abdullaev *et al.* [I.11] put forward methods allowing the description of soliton dynamics in nearly integrable systems.

The selection of material is largely due to the author's interests, and does not claim to give a full description of such an intensively developing field as soliton theory in inhomogeneous media. It is to be noted that a number of aspects of the problems under consideration in the monograph have been put in progress quite recently, so here only the initial steps are described. Especially, this has to do with statistical perturbation theory for solitons (Chapter 4), and the theory of dynamical soliton stochastization (Chapter 5). Further developments are presented in [I.12], [I.13].

The monograph is organized as follows.

Chapter 1 gives asymptotic perturbation theory methods for solitons and nonlinear periodic waves, to be applied to the description of nonlinear wave evolution in inhomogeneous media.

Chapter 2 treats adiabatic dynamics of solitons in inhomogeneous and active media: amplification and damping, resonant motion in varying fields, soliton interaction, collective behavior.

Chapter 3 is dedicated to the description of wave radiation by solitons.

Chapter 4 examines dynamics of nonlinear waves in randomly inhomogeneous media. Chapter 5 studies processes of dynamical chaos of solitons and breathers.

Many scientists from FSU have contributed in this book. It is not possible here to note all their names. My particular thanks to my friends and colleagues from Tashkent – A. A. Abdumalikov, S. A. Darmanyan, B. A. Umarov for fruitful collaborations. Miss Anya Yudine is especially acknowledged for transforming the MS to a readable text.

I am grateful to my family and wife Gulsek for their patience and encouragement. My thanks also to my parents Munik and Khabibulla for their constant support and their years of understanding of my interest in science.

1

PRINCIPLES OF SOLITON THEORY

This chapter, as an introductory one, describes basic representations and results of soliton theory that are presently applied to the investigation of essentially nonlinear systems in condensed matter physics. Let us recall their solutions for a wide class of nonlinear wave propagation and interaction problems can be conventionally brought into two classes: (a) integrable and (b) nonintegrable.

The first class comprises such known equations as the Korteweg–de Vries (KdV), nonlinear Schrödinger (NLS), sine–Gordon (SG), Landau–Lifshitz, Kadomtsev–Petviashvili equations and some others (see monographs [1.1], [1.2], wherein a general soliton theory is fully expounded and many integrable nonlinear wave equations are analyzed).

The nonlinear Klein–Gordon equation with cubic nonlinearity (so-called φ^4-theory), the double sine–Gordon equation, many multidimensional versions of nonlinear wave equations and some others, refer to the class of nonintegrable equations. They have solitary wave solutions. In contrast to the integrable case the interaction of solitary waves is nonelastic. In particular, during their collisons the emission of linear waves takes place. Finally, a wide class of nonintegrable equations is built up from those that can be treated as weak perturbations of completely integrable systems. Such equations are described by soliton propagation in media with weak inhomogeneities of the parameters.

We will often have to deal with them while describing real physical bounded systems and parameter nonuniformities of both deterministic and random nature.

Below we report the main results of soliton theory which are essential for understanding the contents of the monograph. The scheme of the description is built up as follows. Formulae yielding particular solutions as localized nonlinear waves—solitons and stationary nonlinear periodic waves (Sections 1.1–1.7) are derived for all the equations under consideration. Then solutions for two-soliton states—bions and breathers—are written out. Further, the perturbation theory results for nonlinear wave equations as weak perturbations of integrable equations and for perturbed nonintegrable systems are described.

Direct perturbation theory methods and the IST-based methods are described in Sections 1.8–1.9. Asymptotic methods used in the investigation of nonlinear periodic wave evolution are reported in Sections 1.10–1.11.

1.1 THE KORTEWEG–DE VRIES EQUATION

The Korteweg–de Vries equation occurs when describing nonlinear wave processes in many physical problems—hydrodynamics, plasma physics, solid state physics and so on. The derivation of the KdV for concrete problems will not be given here. A detailed derivation is given in monographs by Karpman [1.3], Lamb [1.5] and some special cases are examined in Chapter 4 of the present monograph. The KdV equation (like a number of other equations to be treated below) is as universal (and also as much studied) as the heat-conduction equation—a linear equation of mathematical physics. It describes wave propagation in nonlinear dispersive media.

The dispersion law for the linear limit takes the form

$$\omega = c_0 q - \beta q^3.$$

The medium exhibits nonlinearity of hydrodynamical type, namely, of the type of uu_x, where the subindex denotes a partial derivative with respect to a coordinate.

The KdV equation is

$$u_t - c_0 u_x + \beta u_{xxx} + \alpha u_x u = 0.$$

Proceeding to a reference frame moving with the velocity c_0, and inserting dimensionless variables we will write down the KdV equation in the canonical form

$$u_t - 6uu_x + u_{xxx} = 0. \tag{1.1}$$

The factor 6 has been incorporated into the nonlinear term to simplify the shape of numerical coefficients in the final solutions.

First of all we will analyze general properties of steady state KdV solutions—solitary stationary waves (solitons and multisoliton states) as well as nonlinear periodic solutions. We will try steady state solutions to this equation as travelling waves, $u(x,t)=u(\xi)=u(x-vt)$; $u(\xi)$ is inserted into (1.1) to give

$$-vu_\xi + u_{\xi\xi\xi} - 6uu_\xi = 0.$$

Single integration yields

$$-vu - 3u^2 + u_{\xi\xi} = C_1, \tag{1.2}$$

where C_1 is an integration constant. This equation is consistent in form with that for nonlinear oscillator vibrations.

Both sides of (1.2) are multiplied by u_ξ, to be again integrated to yield

$$-\frac{v}{2}u^2 - u^3 + \tfrac{1}{2}(u_\xi)^2 = C_1 u + C_2, \tag{1.3}$$

where C_2 is the second integration constant. From this we derive

$$(u_\xi)^2 = P(u) = 2C_1 u + 2C_2 + 2u^3 + vu^2. \tag{1.4}$$

It follows from (1.2) that the potential energy of stationary waves is

$$E = -\left(\frac{vu^2}{2} + u^3 + C_1 u\right)$$

The constant C_1 may be removed by a suitable transformation

$$u \to u + \delta,$$

and therefore C_1 is assumed to be zero. A plot of the dependence $E(\varphi)$ and a phase portrait are shown in Figure 1.[1] Here we use the charge $\varphi = -u$. The motion on the separatrix ($C_2 = 0$) in this figure refers to solitons; the orbits that are inside stand for solutions as nonlinear periodic waves. As shown below, near the separatrix periodic waves are described by infinite periodic soliton sequences—soliton lattice. Oscillations about the point A are of harmonic nature.

The point O is a special point, of 'saddle' type, the point A is a point of 'center' type. Solutions describing the motion on the separatrix and near it are found explicitly from the integration of (1.4). Expressing u via φ and writing down a cubic polynomial as the product, we come to

$$(\varphi_\xi)^2 = -(\varphi - \alpha)(\varphi - \beta)(\varphi - \gamma). \tag{1.5}$$

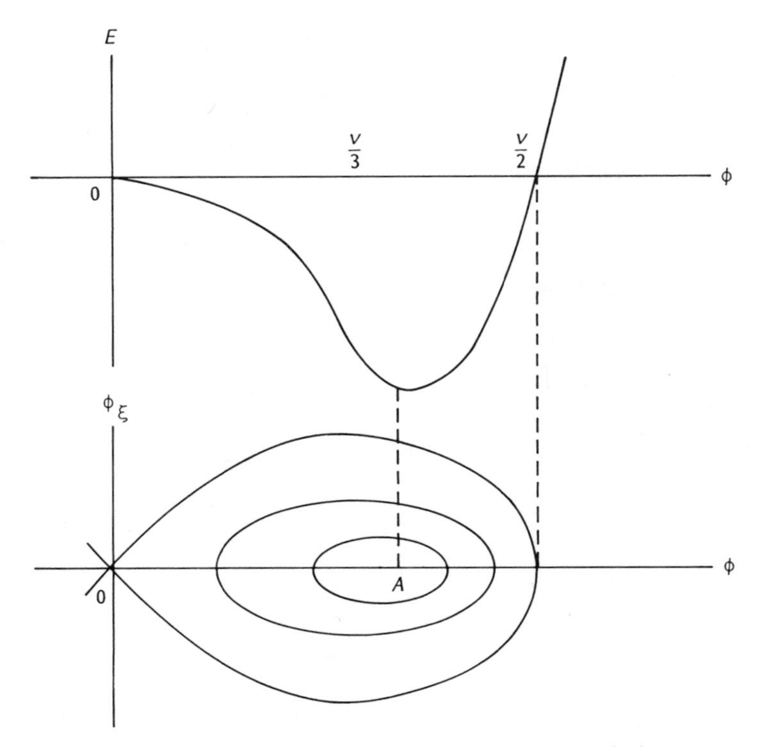

Figure 1.1. Phase portrait of the KdV stationary solutions

Due to real φ_ξ and the real coefficients, the polynomial $P(\varphi)$ has different real roots. From (1.5) we find that

$$\int_{\varphi_0}^{\varphi} \frac{d\varphi}{\sqrt{P(\varphi)}} = \pm\,\xi. \tag{1.6}$$

To define $\varphi(x, t)$ one should solve the problem of inverting the integral in (1.6). This is done with the aid of a Jacobian elliptic function [1.5], to give

$$\varphi(\xi) = \gamma - \mathrm{sn}^2\,[(\gamma-\alpha)\xi, k](\gamma-\beta),$$

$$k^2 = \frac{\gamma-\beta}{\gamma-\alpha}. \tag{1.6a}$$

The case $C_1 = C_2 = 0$ pertains to soliton. Then $\alpha = \beta = 0$ and

$$P(\varphi) = \varphi^2(\gamma - \varphi).$$

For $k \to 1$ we have $\operatorname{sn}(x, k) = \tanh x$. Taking into account the properties of cubic polynomial roots, $\alpha + \gamma + \beta = -v = \gamma$, we see that the solution on the separatrix has the form

$$u = -\varphi = -\frac{v}{2}\operatorname{sech}^2\left[\sqrt{\frac{v}{2}}(x - vt - x_0)\right]. \tag{1.7}$$

Let us write this solution in another form:

$$u_s = -2\kappa^2 \operatorname{sech}^2 \kappa(x - \xi), \qquad \xi = 4\kappa^2 t - \xi_0, \quad \kappa = \sqrt{\frac{v}{2}}. \tag{1.8}$$

Solitons of larger amplitudes are seen to move faster. The soliton width is $\Delta = \kappa^{-1}$, so the relations

$$\frac{\Delta^2 A}{2} = 1, \qquad A = 2\kappa^2$$

are valid.

Note the following important property of the solution (1.8). The KdV equation shows that in the soliton solution, contributions of the dispersive and nonlinear terms are equal.
This reflects the fundamental fact that nonlinear stationary waves occur *provided* the influence of nonlinear effects in the system is balanced by dispersive effects.

Let us evaluate the soliton energy. Taking (1.8) into consideration, we obtain

$$E_s = -\int_{-\infty}^{\infty} \mathrm{d}x(u^3 + \tfrac{1}{2}u_x^2) = \tfrac{32}{5}\kappa^5. \tag{1.9}$$

For arbitrary α, β, γ the solution (1.6a) oscillates between the values $u_1 = \beta$ and $u_2 = \gamma$, with period $T = 2K/(\gamma - \alpha)$ where $K(k)$ is the complete elliptic integral of the first kind,

$$K(k) = \int_0^{\pi} \frac{\mathrm{d}\varphi}{\sqrt{(1 - k^2 \sin^2 \varphi)}}.$$

For $\gamma \to \beta$ and $k \ll 1$ we have $\operatorname{sn}(x, k) \to \sin x$, and the solution takes the form of a monochromatic wave.

At the present time the KdV equation has been studied in detail. The IST allows the solution of the problem for a randomly decreasing initial state, the determination of the asymptotic behavior of solutions for long times, the solution of the periodic initial problem and others. A detailded analysis of these problems is presented in monographs [1.1–1.4]. Here we only report IST results for two-soliton complexes and soliton evolution under perturbation, within the adiabatic perturbation.

A two-soliton solution has the form

$$u(x,t) = -(2\kappa_2^2 - 2\kappa_1^2)\frac{\kappa_2^2 \operatorname{cosech}^2 \gamma_2 + \kappa_1^2 \operatorname{cosech}^2 \gamma_1}{(\kappa_2^2 \coth^2 \gamma_2 + \kappa_1 \coth \gamma_1)}, \tag{1.10}$$

$$\gamma_{1,2} = \kappa_{1,2}x - 4\kappa_{1,2}^3 t + \delta_{1,2},$$

where $\delta_{1,2}$ are integration constants (the soliton phase shift). Consideration of this formula shows that for times much longer than those of soliton interactions, this solution can be treated as a system of two solitons with slightly overlapping tails.

1.2 THE NONLINEAR SCHRÖDINGER EQUATION (NLSE)

The NLSE, along with the KdV, is one of the universal nonlinear wave equations that is widely applied to the solution of different problems in plasma physics, condensed matter physics, superconductivity, etc. This equation is the first one to which the IST has been applied [1.6] and its complete integrability has been proved.

The NLS in the canonical form is

$$\mathrm{i}q_t + q_{xx} + 2|q|^2 q = 0. \tag{1.10}$$

A solution of this equation is sought in the form

$$q(x,t) = A(x - vt)\mathrm{e}^{\mathrm{i}\varphi t + \mathrm{i}\psi x}. \tag{1.11}$$

Substitution of (1.11) into (1.10) yields

$$A_{\xi\xi} + \mathrm{i}(2\psi - v)A_\xi - A(\psi + \varphi^2) + 2A^3 = 0, \quad \xi = x - vt. \tag{1.12}$$

Let the complex coefficient before A_ξ vanish. Then

$$\psi = \frac{v}{2}.$$

Further, from the equality

$$\varphi = -\psi^2 + B$$

we find that

$$\varphi = -\frac{v^2}{4} + B.$$

Finally from (1.12) we obtain

$$A_{\xi\xi} - BA + 2A^3 = 0 \tag{1.13}$$

The last equality is integrated, to result in

$$(A_\xi)^2 = C + BA^2 - A^4 \tag{1.14}$$

Then the solution for A is found by inverting an elliptic integral,

$$\int_{A_0}^{A} \frac{\mathrm{d}A}{\sqrt{(C + BA^2 - A^4)}} = \pm \xi. \tag{1.15}$$

Let the quartic polynomial occurring here be represented as

$$P(A) = (\alpha_1 - A^2)(A^2 - \alpha_2).$$

Then we have, with $F(A, K)$ the incomplete elliptic integral,

$$\frac{1}{\sqrt{\alpha_1}}\{K(k) - F(A, k)\} = \pm \xi, \quad k = \sqrt{\left(\frac{\alpha_1 - \alpha_2}{\alpha_1}\right)},$$

$$\alpha_{1,2} = \frac{B}{2} \pm \sqrt{\left(\frac{B^2}{4} + C\right)}, \qquad \alpha_{1,2} = A_{1,2}^2.$$

From this we have

$$A(\xi) = A_1\left\{1 - \left[\left(1 - \frac{A_1^2}{A_2^2}\right)\mathrm{sn}^2(\xi, k)\right]\right\}^{1/2}. \tag{1.16}$$

For $C \to 0$, $A_1 \to A_0$, $k \to 1$,

$$A \to A_0 \operatorname{sech} A_0 \xi, \qquad A_0 = \sqrt{B}.$$

Ultimately the soliton solution is

$$q_s(x, t) = \frac{\sqrt{B}}{\cosh \sqrt{B}(x - vt)} \exp\left\{-\mathrm{i}\frac{v^2}{4}t + \frac{\mathrm{i}vx}{2} + \mathrm{i}Bt\right\}.$$

Insertion of $\sqrt{B} = 2\eta$, $v = -4\xi$ into the above expression leads us to a canonical form for the one-soliton solution

$$q_s(x,t) = 2\eta \operatorname{sech} 2\eta(x + 4\xi t - x_0) \exp\{-2i\xi x + 4i(\eta^2 - \xi^2)t + i\delta_0\}. \tag{1.17}$$

Note that unlike the KdV soliton, the NLS soliton amplitude is independent of its velocity. The NLS soliton describes a modulated wave packet propagating in a nonlinear dispersive medium with a constant velocity (Figure 1.2).

The dispersion law corresponding to (1.10) is

$$\omega = k^2 - 2A_0^2.$$

Let us evaluate the energy for the NLS soliton. The NLS Hamiltonian is

$$H = \int_{-\infty}^{\infty} (|q_x|^2 - |q|^4)\, dx. \tag{1.18}$$

For a single soliton the integration gives

$$H = 32\eta(\xi^2 - \tfrac{1}{3}\eta^2). \tag{1.19}$$

In a similar manner one can evaluate the number of particles N and the

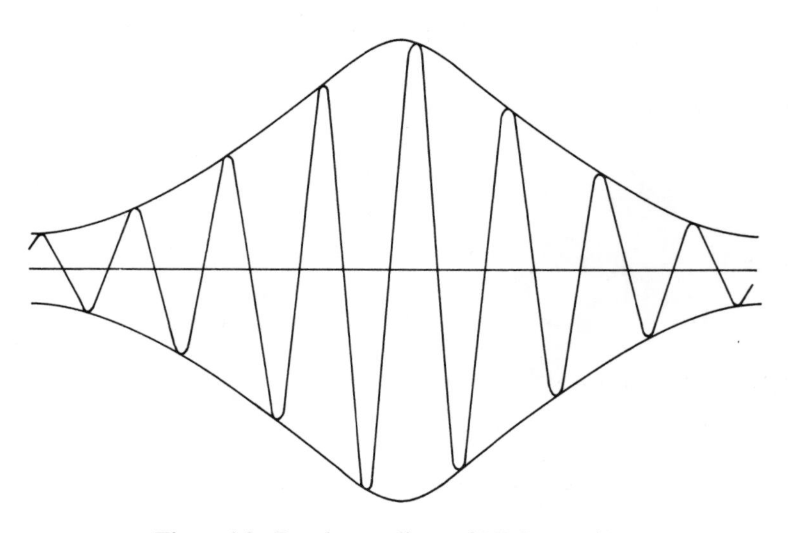

Figure 1.2. Envelope soliton of NLS equation

field momentum P for a single soliton:

$$N \equiv \int_{-\infty}^{\infty} |q|^2 \, dx = 4\eta \tag{1.20}$$

$$P \equiv -\frac{i}{2}\int_{-\infty}^{\infty} (q^* q_x - q_x^* q) \, dx = 8i\xi\eta. \tag{1.21}$$

Now let us write out (with no algebra) a formula describing a two-soliton solution for NLSE:

$$q_B(x, t) = 4\exp\left(-\frac{it}{2}\right) \cdot \frac{\cosh(3x) + 3\exp(-4it)\cosh(x)}{\cosh(4x) + 4\cosh(x) + 3\cos(4t)}, \qquad t' = t/2. \tag{1.22}$$

At the points of maximum $|q|$ we have $|q| \sim 1$. The oscillation period of this solution is $T/2 = \pi/2$.

1.3 THE SINE–GORDON EQUATION

Unlike the KdV and NLSE equations, this equation is of the second order in time,

$$\varphi_{tt} - c_0^2 \varphi_{xx} + \omega_0^2 \sin\varphi = 0, \tag{1.23}$$

where c_0 is the limiting velocity and ω_0 is a characteristic frequency.

Now we will proceed with the dimensionless variables

$$t = \omega_0 t, \qquad x = \frac{x}{d}, \qquad d = \frac{c_0}{\omega_0}.$$

The SG equation in the canonical form is

$$\varphi_{tt} - \varphi_{xx} + \sin\varphi = 0. \tag{1.24}$$

Steady state solutions are sought in the form $\varphi = \varphi(x - vt)$. Substituting $\varphi(\xi)$ into (1.3.2), multiplying both sides by φ_ξ and integrating, we obtain an ordinary differential equation

$$\tfrac{1}{2}(v^2 - 1)\varphi_\xi^2 + 1 - \cos\varphi = H. \tag{1.25}$$

Its phase portrait is depicted in Figure 1.3 (for waves with $v^2 - 1 > 0$).

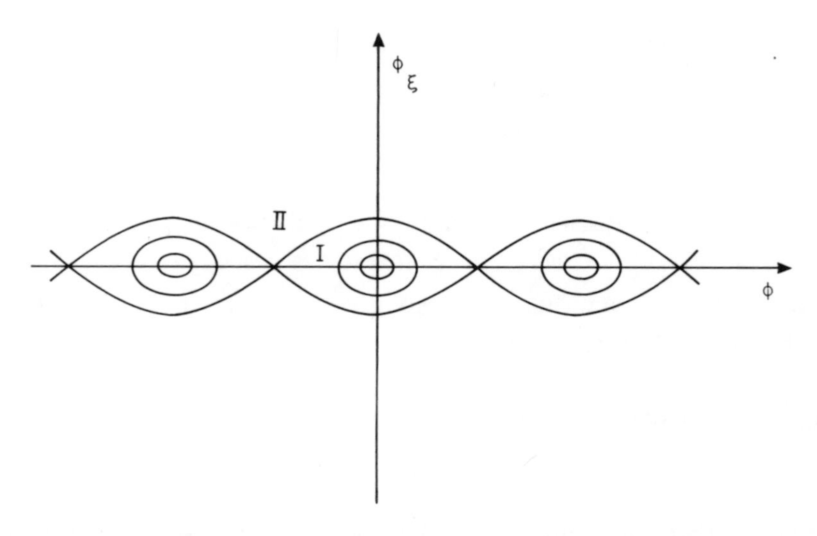

Figure 1.3. Phase-plane portrait for the SGE stationary waves (when $v^2 - 1 > 0$)

Stationary waves are distinguished by the values for phase velocity $V > 1$ (fast waves), $V < 1$ (slow waves) and period-averaged derivatives $\langle \varphi_t \rangle$ and $\langle \varphi_x \rangle$. In the region I $\langle \varphi_t \rangle = \langle \varphi_x \rangle = 0$, and in the region II either $\langle \varphi_t \rangle$ or $\langle \varphi_x \rangle$ is nonzero. The regions of different values for H correspond to different types of wave motions: $H = 0$ stands for solitons (kinks), describing a 2π turnover, and $H \neq 0$ refers to nonlinear periodic and spiral waves.

Let us inspect the cases with different values of $v^2 - 1$ and H:

(a) $v^2 - 1 < 0$, $H = 0$. From (1.25) we have

$$\int_{\varphi_0/2}^{\varphi/2} \frac{\mathrm{d}\varphi}{\sin \varphi} = \pm \frac{\xi}{\sqrt{(1 - v^2)}},$$

from which we find

$$\varphi = 4 \arctan \left\{ \exp\left[\pm \frac{(x - vt - x_0)}{\sqrt{(1 - v^2)}} \right] \right\}. \tag{1.26}$$

The sign $(+)$ pertains to the soliton (kink) and $(-)$ to the antisoliton (antikink) (Figure 1.4 for kink).

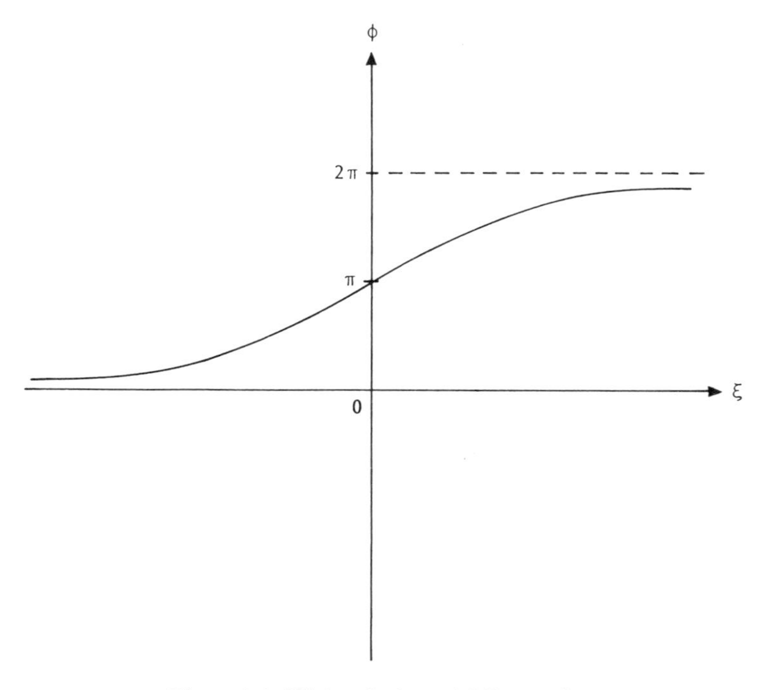

Figure 1.4. Kink solution of SG equation

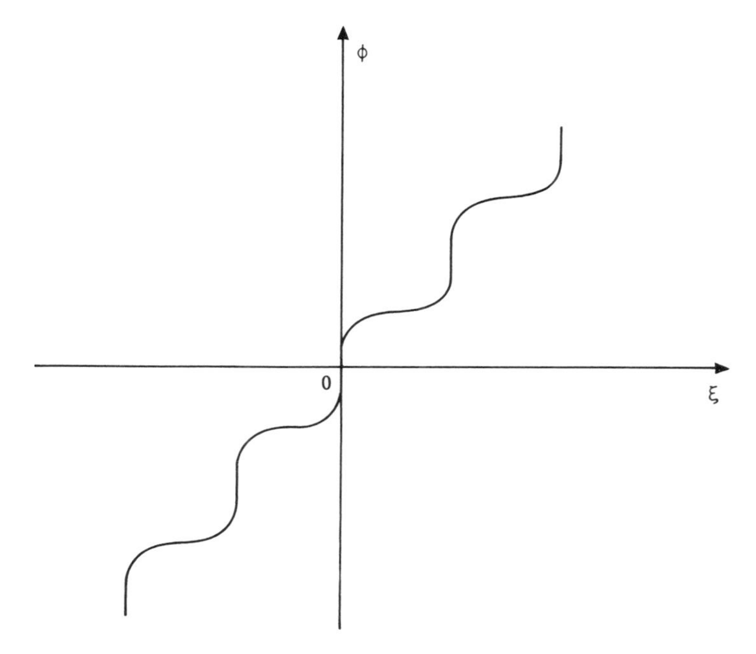

Figure 1.5. Nonlinear periodic wave solution of SGE

(b) $v^2 - 1 < 0$, $H < 0$. Eq. (1.3.3) is integrated to give

$$\int_{\varphi_0}^{\varphi} \frac{d\varphi}{\sqrt{\left(|H| + 2\sin^2\frac{\varphi}{2}\right)}} = \pm\frac{\sqrt{2}\xi}{\sqrt{(1-v^2)}}, \quad \varphi_0 = \pi.$$

From this we have

$$\varphi = \pi + 2\,\mathrm{am}\left[\pm\frac{\xi}{\sqrt{(1-v^2)}}\left(1+\frac{H}{2}\right)^{1/2}\right]. \tag{1.27}$$

The function am x is an elliptic (Jacobi) amplitude. Its plot is depicted in Figure 1.5.

The resulting solution is consistent with rotations of $\varphi(x)$ through 2π-fold angles. In the theory of Josephson functions, quantities of physical meaning are the phase derivatives $\varphi_t \sim E$ and $\varphi_x \sim H_0$, corresponding to electric and magnetic fields. For the magnetic field we have

$$H_0 = 2\sqrt{\left[\frac{H}{2} + \mathrm{cn}^2\left(y\sqrt{1+\frac{H}{2}}\right)\right]},$$

$$x = \frac{\sqrt{(2\xi)}}{\sqrt{(1-v^2)}}, \qquad \mathrm{cn}(y) = \cos(amx).$$

This expression describes a chain of magnetic vortices (soliton lattice) in a long Josephson junction.

One can obtain another form for the solution $\varphi(x, t)$ if the definitions

$$\varphi = am\,u, \quad \sin\varphi = \mathrm{sn}\,u = x$$

are taken into account. Then from (1.27) we derive

$$\varphi = \pi + 2\arcsin\left[\mathrm{sn}\left(x\sqrt{1+\frac{|H|}{2}}, k\right)\right],$$

$$\frac{|H|}{2} = \frac{1-k^2}{k^2}.$$

In a similar manner we build up solutions for other values of $v^2 - 1$ and H (see also [1.7], [1.8]).

Now let us construct the two-soliton solution. There are two possibilities; the first one is a system of two interacting solitons of the same polarity (bions), the second one is a system of two interacting solitons of different polarity (breathers).

Here we write out the IST formula for an SG breather:

$$\varphi_{\mathrm{B}} = 4\tan^{-1}\left\{\frac{\eta}{|\nu|}\frac{\sin\left[\dfrac{\nu}{|\lambda|}\left(\dfrac{t-vx}{\sqrt{(1-v^2)}}-\theta_0\right)\right]}{\cosh\left[\dfrac{\eta}{|\lambda|}\left(\dfrac{x-vt-x_0}{\sqrt{(1-v^2)}}\right)\right]}\right\}, \tag{1.28}$$

$$\lambda = \nu + \mathrm{i}\eta, \quad \lambda \text{ is the spectral parameter,}$$

where v is the breather velocity, equal to

$$v = \frac{4|\lambda|^2-1}{4|\lambda|^2+1}, \qquad \text{with} \quad x_0 = \frac{1}{2\,\mathrm{Im}\,\lambda}\ln\left[\frac{\nu}{\eta}\frac{c}{2\lambda}\right],$$

$$\theta_0 = \frac{|\lambda|}{\nu}\varphi_0, \qquad \varphi_0 = \arg\left(\frac{c}{2\lambda}\right).$$

($\mathrm{d}\theta_0/\mathrm{d}t = \omega_{\mathrm{B}}$ is the frequency of intrinsic breather pulsations). Note that an SG soliton (kink) is a *topological* soliton. It interpolates two different ground states (or two different vacua in the quantum theory formalism). For them the value of the topological charge Q is nonzero:

$$Q = \frac{1}{2\pi}\int_{-\infty}^{\infty}\varphi_{\mathrm{s}}\,\mathrm{d}x = 1.$$

For a breather $Q = 0$. On the other hand a KdV soliton is a dynamical soliton and may be eliminated by continuous deformation, under the action of dissipation and other mechanisms.

Let us evaluate the energy of the kink and breather. For the kink we have

$$E_{\mathrm{k}} = \tfrac{1}{2}\int_{-\infty}^{\infty}\mathrm{d}x(\varphi_t^2 + \varphi_x^2 + 4\sin^2(\varphi/2)) = \frac{8}{\sqrt{(1-v^2)}}.$$

For the breather,

$$E_{\mathrm{B}} = 16\sin\gamma(1-v^2)^{-1/2}, \qquad \gamma = \tan^{-1}\left(\frac{\nu}{\eta}\right).$$

The threshold kink energy is $E = 8(v = 0)$, whereas the breather energy starts from zero. For completeness we write down a remarkable formula for the SG Hamiltonian derived by Faddeev and Kovepin [1.9]:

$$H_{\mathrm{SG}} = \int_0^\infty \left(\frac{1}{4\lambda} + \lambda\right) P_\lambda \, \mathrm{d}\lambda + \sum_k \frac{2}{\lambda}(4\,\mathrm{e}^{-P_k} + 4\,\mathrm{e}^{P_k})$$

$$+ \sum_k \frac{4}{\gamma} \sin\frac{\gamma\eta_k}{16}(4\,\mathrm{e}^{-\eta_k/4} + 4\,\mathrm{e}^{\eta_k/4}).$$

The contributions of the continuous spectrum, solitons and breathers to the energy are expressed by the first, second and third terms respectively, in this formula. This form of representation appears useful for calculations of the energy of waves radiated by solitons and breathers in inhomogeneous and transient media.

Now we will analyze the breather in more detail. The case $\gamma \ll 1$ corresponds to a small-amplitude breather,

$$\varphi_{\mathrm{B}} \approx \frac{\gamma \sin t}{\cosh \gamma x}. \tag{1.29}$$

For $\gamma \to \pi/2$ the breather can be treated as a system weakly interacting, kink and antikink. As shown in [1.10], the breather can be characterized by the relative distance $r = x_2 - x_1$ between kink and antikink, the equation for which is of the form

$$\frac{\mathrm{d}^2 r}{\mathrm{d}t^2} = -8r \exp(-|r|), \tag{1.30}$$

and the energy is

$$E = \frac{p^2}{2} + U(r), \qquad U(r) = -8\,\mathrm{e}^{-|r|},$$

i.e. equation (1.30) coincides with that of the motion of unit mass particles in the exponential potential of attraction.

Indeed, following [1.11], we have for a kink–antikinik, separated by distance q

$$\varphi(x,t) = \psi(x,t) + \varphi(x - vt + q) + \bar{\varphi}(x + vt - q)$$

$$\psi_{tt}|_{t=0} = \sum_i \sin\varphi_i - \sin\left(\sum_i \varphi_i\right) \simeq \frac{-8\rho^{-q}}{\cosh^2(x-q)} \qquad (q \gg 1, x \sim q).$$

The breather Hamiltonian, in terms of the variables γ and θ, is

$$H = \sin\gamma.$$

Figure 1.6. Phase portrait of breather in γ, θ plane

Figure 1.6 shows its phase portrait in the γ, θ plane. For $\gamma > 1/2$ the breather disintegrates into free kink and antikink parts.

1.4 THE LANDAU–LIFSHITZ EQUATION

Let us consider an isotropic ferromagnetic. Dynamics of the magnetization vector **S** are described by the Landau–Lifshitz (L–L) equation

$$\frac{\partial \mathbf{S}}{\partial t} = \tfrac{1}{2} I \mathbf{S}(x, t) \times \nabla^2 \mathbf{S}(x, t). \tag{1.31}$$

Lakshmanan [1.12] and Takhtajan [1.13] showed that the one-dimensional version of the L–L equation is an integrable system that can be reduced to the NLS. Below we shall report results on proving equivalence of the L–L equation with the NLS. A detailed description of the magnetic soliton theory, different solitons and N-soliton solutions of the L–L equation is given in the monograph by Kosevich *et al.* [1.14].

The one-dimensional version of the L–L equation is [1.15]:

$$\frac{\partial \mathbf{S}(x, t)}{\partial t} = \mathbf{S} \times \frac{\partial^2 \mathbf{S}}{\partial x^2}. \tag{1.32}$$

The spin energy density is

$$E(x,t)=\frac{1}{2}\left|\frac{\partial \mathbf{S}}{\partial x}\right|^2, \tag{1.33}$$

and the current density is

$$j=\mathbf{S}\cdot\left(\frac{\partial \mathbf{S}}{\partial x}\times\frac{\partial^2 \mathbf{S}}{\partial x^2}\right). \tag{1.34}$$

The magnitudes E and j satisfy the continuity equation

$$\frac{\partial E}{\partial t}+\frac{\partial j}{\partial x}=0.$$

To find soliton solutions of (1.32) we will proceed with the angular variables $\theta(x,t)$ and $\varphi(x,t)$ for the unit vector $\mathbf{S}$. Then the expression (1.32) is equivalent to a set of equations

$$\theta_t=-2\theta_x\varphi_x\cos\theta-\varphi_{xx}\sin\theta \tag{1.35}$$

$$\sin\theta\,\varphi_t=\theta_{xx}-(\varphi_x)^2\sin\theta\cos\theta. \tag{1.36}$$

Further, we have equalities

$$E=\tfrac{1}{2}[(\theta_x)^2+\chi^2], \qquad \chi=\sin\theta\,\varphi_x.$$

A solution of (1.35) and (1.36) is tried in the form

$$\theta=\theta(x-vt), \qquad \varphi(x,t)=\bar{\varphi}(x-vt)+\Omega t.$$

Then we have

$$\chi=\varphi_x\sin\theta,$$

$$\chi_\xi=\theta_\xi(v-\chi\cot\theta), \tag{1.37}$$

$$\theta_{\xi\xi}=-\chi(v-\chi\cot\theta)+\Omega\sin\theta. \tag{1.38}$$

Hence

$$(\theta_\xi)^2+\chi^2+2\Omega\cos\theta=2\alpha, \qquad E=-\Omega\cos\theta+\alpha, \qquad \alpha=\text{const.} \tag{1.39}$$

It is readily seen that

$$\frac{\mathrm{d}\varphi}{\mathrm{d}\xi}=\frac{v}{1+\tanh^2\left(\dfrac{v\xi}{2}\right)},$$

$$\varphi(x,t) = \arctan\left[\tanh\frac{v}{2}(x-vt)\right] + \tfrac{1}{2}vx, \tag{1.40}$$

$$\cos\theta = \left(\tanh\frac{v\xi}{2}\right)^2, \qquad \Omega = \frac{v^2}{2},$$

$$E(\xi) = \frac{v^2}{2\cosh^2\left(\dfrac{v\xi}{2}\right)}, \qquad j = cE(\xi). \tag{1.41}$$

Now we will show that equation (1.32) is equivalent to the NLSE [1.12]. In this case we will progress from **S** to the Serret–Frenet vectors. Let us denote curvature by κ torsion by τ. For the energy and momentum there are relations

$$E(x,t) = \tfrac{1}{2}\kappa^2, \qquad j(x,t) = \kappa^2(x,t)\tau(x,t).$$

The continuity equation becomes

$$\kappa_t = -2\kappa_x\tau - \kappa\tau_x, \tag{1.42}$$

and the compatibility conditions yield

$$\tau_t = \left(\frac{\kappa_{xx}}{\kappa} - \tau^2\right)_x + \kappa\kappa_x. \tag{1.43}$$

Now let us introduce the function

$$\psi = \kappa(x,t)\exp\left\{\mathrm{i}\int^x \tau(x,t)\,\mathrm{d}x\right\}. \tag{1.44}$$

Then equations (1.42) and (1.43) turn into the NLSE:

$$\frac{1}{\mathrm{i}}\psi_t = \psi_{xx} + \tfrac{1}{2}|\psi|^2\psi, \tag{1.45}$$

which has been already solved (see Section 1.2).

For the single-soliton solution we obtain

$$E(x,t) = \tfrac{1}{2}\kappa^2 = 8\eta^2\operatorname{sech}^2\{2\eta(x-x_0) + 8\eta\xi t\}$$
$$j(x,t) = \kappa^2\tau = 32\xi\eta^2\operatorname{sech}^2\{2\eta(x-x_0) + 8\eta\xi t\},$$

while coincides with the solution (1.40) provided

$$\tau = -2\xi = \frac{v}{2}, \qquad \kappa = 4\eta.$$

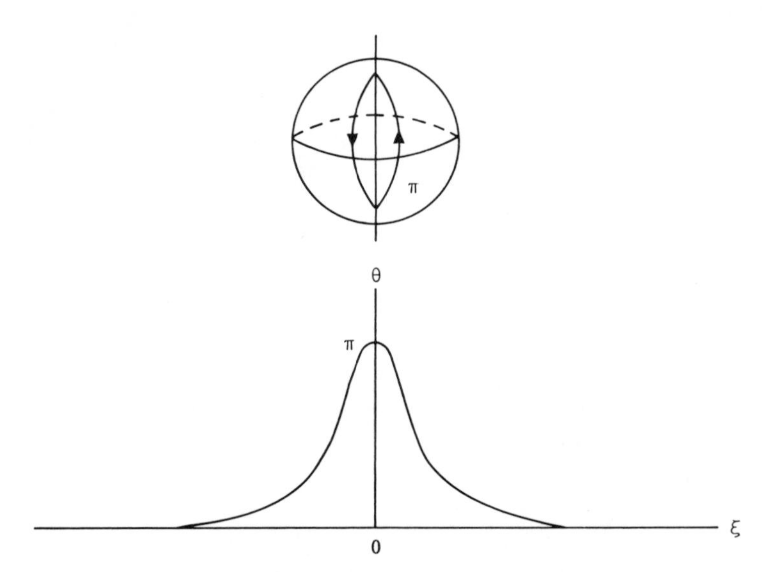

Figure 1.7. Turn of magnetization and its distribution in the Landau–Lifshitz equation solitons

Plots describing the turning of magnetization and its distribution in solitons are shown in Figure 17(a, b).

1.5 THE KLEIN–GORDON NONLINEAR EQUATION

Now we will investigate the Klein–Gordon cubic nonlinear equation (φ^4 model) in the form

$$\varphi_{tt} - \varphi_{xx} + \varphi^3 - \varphi = 0. \tag{1.46}$$

The corresponding Lagrangian density is

$$\mathscr{L} = \tfrac{1}{2}\left\{\varphi_t^2 - \varphi_x^2 + \varphi^2 - \frac{\varphi^4}{2}\right\}. \tag{1.47}$$

In contrast to the previously studied equations the φ^4 theory is non-integrable. Equation (1.46) appears in problems of quantum field theory [1.16], in the theory of structural phase transitions [1.17], and in many other problems.

Let us find steady-state solutions of this equation. Let $\varphi = \varphi(x - vt)$. From (1.46) we have

$$(v^2 - 1)\varphi_{\xi\xi} - \varphi + \varphi^3 = 0. \tag{1.48}$$

Multiplication of both sides by φ_ξ and its integration yields

$$(v^2-1)(\varphi_\xi)^2-\varphi^2+\tfrac{1}{2}\varphi^4=C. \tag{1.49}$$

$C(\varphi)$ is minimal for $C=-\frac{1}{2}$. In this case we have

$$\int_{\varphi_0}^{\varphi}\frac{\mathrm{d}\varphi}{\varphi^2-1}=\pm\frac{\xi}{\sqrt{[2(1-v^2)]}}.$$

Hence we have

$$\varphi(x,t)=\pm\tanh\left[\frac{x-vt-x_0}{\sqrt{[2(1-v^2)]}}\right]. \tag{1.50}$$

It is seen that $|\varphi|\to 1$ for $|\xi|\to\infty$. The signs $(\pm)$ refer to two different types of 'solitons' (kink and antikink).

Let us determine the energy of a moving 'soliton' (kink):

$$E=\frac{1}{2}\int_{-\infty}^{\infty}\left\{\varphi_t^2+\varphi_x^2-\varphi^2+\frac{\varphi^4}{2}+\frac{1}{2}\right\}\mathrm{d}x$$

$$=\frac{2\sqrt{2}}{3}\frac{1}{\sqrt{(1-v^2)}}\simeq E_0+m^*v^2/2,\quad E_0\simeq\frac{2\sqrt{2}}{3},\quad V\ll 1. \tag{1.51}$$

The rest mass and momentum p of the kink are

$$m^*=\frac{2\sqrt{2}}{3},\quad p=\int_{-\infty}^{\infty}\mathrm{d}x\varphi_t\varphi_x=\frac{m^*v}{\sqrt{(1-v^2)}}.$$

Now we will write out nonlinear periodic solutions. For $C\neq-\frac{1}{2}$ we have

$$\int_{\varphi_0}^{\varphi}\frac{\mathrm{d}\varphi}{\sqrt{[(\varphi^2-a^2)(\varphi^2-b^2)]}}=\pm\frac{\xi}{\sqrt{[2(1-v^2)]}},$$

$$a=1+\sqrt{(1+2C)},\qquad b=1-\sqrt{(1+2C)}.$$

Inverting the elliptic integral, we find that

$$\varphi(x,t)=a\,\mathrm{sn}\left(\frac{a\xi}{\sqrt{[2(1-v^2)]}},\frac{b}{a}\right). \tag{1.52}$$

The period of the solution is

$$4K(k),\qquad k=b/a.$$

The behavior of the solutions for the 'soliton and nonlinear periodic wave is depicted in Figure 1.8.

The interaction between the kink and antikink proceeds via the exponential attraction potential. As numerical experiments [1.18] show, nonintegrability of (1.46) gives rise to extra features in the kink interaction.

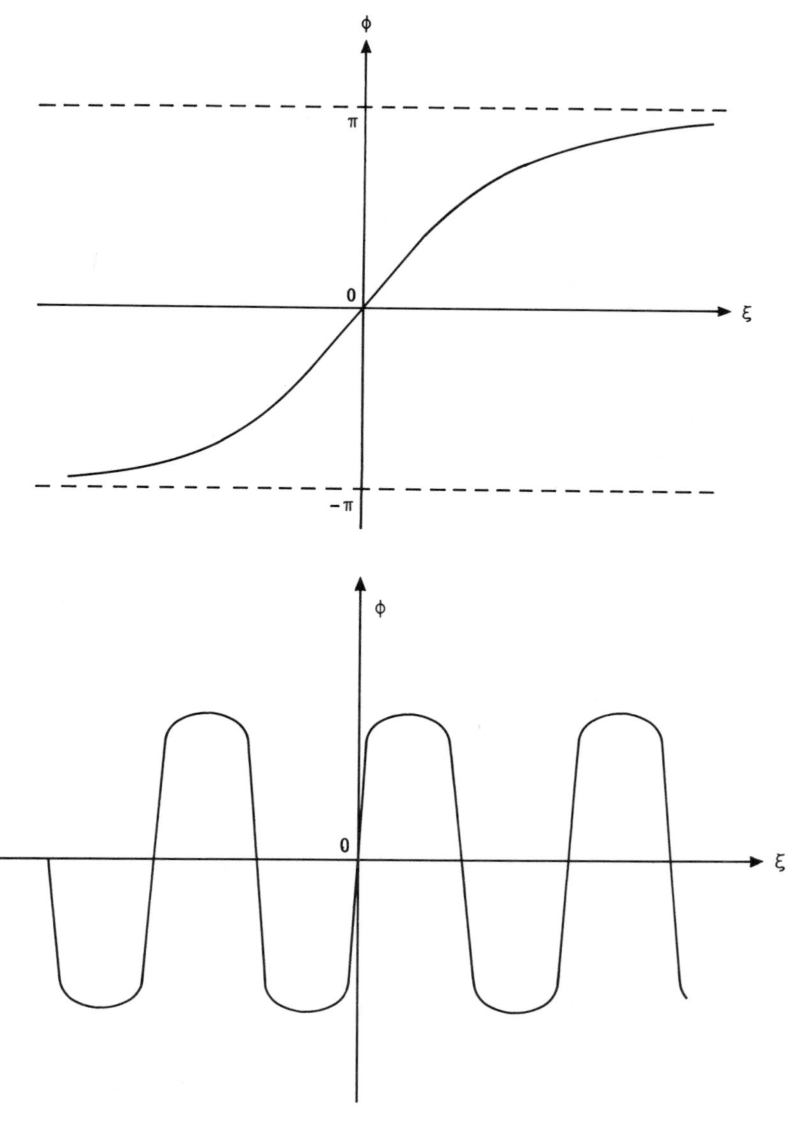

Figure 1.8. Kink and nonlinear periodic wave of the φ^4-model

In integrable systems breathers live indefinitely long, and there is no radiation at all. In φ^4 theory, when kink and antikink collide for velocities lower than a critical one ($v_{\text{crit}} = 0.2$), a time-oscillating bound state is formed, the extra energy being radiated as linear waves. At the same time, though the bion is unstable, its lifetime appears fairly long.

1.6 THE DOUBLE SINE–GORDON EQUATION

Now we will investigate the so-called double sine–Gordon equation (DSG):

$$\varphi_{tt} - \varphi_{xx} + \sin\varphi + 2\lambda \sin 2\varphi = 0. \tag{1.53}$$

This is also of nonintegrable type. Equation (1.53) occurs when one studies dynamics of magnetization in weak ferromagnets, in nonlinear optics, and in other problems. Let us find its steady-state solutions. Going as usual to the variable $\xi = x - vt$, from (1.53) we derive the expression

$$\tfrac{1}{2}(v^2 - 1)(\varphi_\xi)^2 + (1 - \cos\varphi) + \lambda(1 - \cos 2\varphi) = C. \tag{1.54}$$

Let us analyze the case $C = 0$. From (1.54) we find that

$$\int_{\varphi_0}^{\varphi} \frac{\mathrm{d}\varphi}{\sqrt{(1 - \cos\varphi + 2\lambda \sin^2\varphi)}} = \frac{\sqrt{2}\xi}{\sqrt{(1 - v^2)}}. \tag{1.55}$$

Evaluation of the integral in (1.55) gives the solution

$$\varphi(x, t) = 2\tan^{-1}\left\{\sqrt{(1 + 4\lambda)}\sinh^{-1}\left[\frac{\sqrt{(4\lambda + 1)}\xi}{\sqrt{(1 - v^2)}}\right]\right\}. \tag{1.56}$$

For $\lambda \to 0$ this turns into a 2π-soliton solution of the SG equation.

Another form of entry of the DSG equation is

$$\varphi_{tt} - \varphi_{xx} = \pm\left(\sin\varphi + \tfrac{1}{2}\lambda \sin\frac{\varphi}{2}\right). \tag{1.57}$$

The general traveling-wave solution of this equation takes the form

$$\begin{aligned}\varphi = \varphi_0 &+ 4\tan^{-1}[\cosh\delta_3 \exp(\theta - \delta_1) + \sinh\delta_3] \\ &+ 4\tan^{-1}[\cosh\delta_3 \exp(\theta - \delta_3) - \sinh\delta_3],\end{aligned}$$

$$\theta = \omega(t - x/v) + \theta_0 = \frac{\omega}{2\sqrt{(\alpha v)}}[(2v-1)\tau - \xi] + \theta_0,$$

$$\varphi_t = \omega\frac{d\varphi}{d\theta}, \qquad \xi = \sqrt{[\alpha(2x-t)]}, \qquad \tau = \sqrt{(\alpha t)}.$$

As in the case of the φ^4-model, for the DSG equation there are several numerical calculations on kink and antikink collisions. During the collision, waves are radiated, hence in this case one can treat breathers as *metastable states* [1.19]. The energy for the DSG soliton is evaluated by standard means as

$$E_s = 4\sqrt{\left(\frac{1+\beta}{1-v^2}\right)}[1 + J(\beta)],$$

$$J(\beta) = \frac{1}{2\sqrt{[\beta(\beta+1)]}}\ln\left|\frac{1+\beta+\sqrt{[\beta(1+\beta)]}}{1+\beta-\sqrt{[\beta(1+\beta)]}}\right|, \qquad \beta = 4\lambda. \tag{1.58}$$

It is to be noted in the following that the DSG breathers live rather long (many hundreds of periods), and they can be moved by the external fields [1.20]. When the external periodic field is applied, in the equation (1.53) the perturbation $\varepsilon R = \varepsilon \sin \Omega t$ results. The analysis of breather dynamics on the basis of perturbation theory (see Section 1.9) indicates the existence of breathers that are synchronised at frequency ω_B. It can be shown that the loss is compensated at the expense of the external field energy, and the breather states are stabilized. For small λ the equation for relative distance r can be written [1.10]:

$$\frac{d^2 z}{dt^2} = \sigma_1\sigma_2\left(8\,e^{-|r|} + \gamma_1\gamma_2\frac{\lambda}{2}r\right)$$

1.7 A SET OF EQUATIONS DESCRIBING THE INTERACTION BETWEEN HIGH- AND LOW-FREQUENCY WAVES

Let us consider solutions of a set of two coupled equations describing high-and low-frequency wave interaction.

This set of equations occurs in the study of a broad class of phenomena in physics. In particular, the problems of the interaction between electrons and lattice acoustic oscillations (acoustic polarons) [1.21], excitons and phonons [1.22], Langmuir and ion-sound oscillations pertain to this set. The initial set of equations, in terms of dimensionless variables, is written down in the form (so-called Zakharov's system)

$$\mathrm{i}\psi_t + \psi_{xx} - n\psi = 0, \tag{1.59}$$

$$n_{tt} - n_{xx} = (|\psi|^2)_{xx}. \tag{1.60}$$

To find steady-state solutions of this set of equations we represent $\psi(x, t)$ and $n(x, t)$ as

$$n = n(\xi) = n(x - vt), \qquad \psi(x, t) = \varphi(\xi)\exp(\mathrm{i}\alpha x + \mathrm{i}\beta t). \tag{1.61}$$

From (1.60) we derive

$$(v^2 - 1)n_{\xi\xi} + (\varphi^2)_{\xi\xi} = 0.$$

Then after integration over ξ and with the condition $\varphi \to 0$ for $|\xi| \to \infty$, we obtain

$$n(\xi) = -\frac{\varphi^2}{1 - v^2}. \tag{1.62}$$

Substitution of (1.62) into (1.59) yields ($\psi' = \psi/\sqrt{(1 - v^2)}$, $x' = x/v^2$, $t' = t/2$)

$$\mathrm{i}\psi'_{t'} + \psi'_{x'x'} + 2|\psi'|^2\psi' = 0. \tag{1.63}$$

This equation has been solved in Section 1.2. For (1.63) the solution is

$$\psi(x, t) = [2\eta^2(1 - v^2)]^{1/2}\operatorname{sech}(\eta\xi)\exp\left\{\mathrm{i}\left[-\frac{v}{2}\xi + \left(\eta^2 + \frac{v^2}{4}\right)t\right]\right\}.$$

In the theory by Davydov and Kislukha [1.22], $\int_{-\infty}^{\infty} \varphi^2\,\mathrm{d}x = 1$; hence, for $v \to 1$, $n(x, t)$ exhibits a singularity, indicating the need for modification of the initial system (1.59) and (1.60), namely, taking into account dissipation, anharmonicity, etc. It is deduced from (1.62) that the envelope soliton in the high-frequency system is accompanied by a soliton in the low-frequency system, moving with the same velocity.

Equations (1.59) and (1.60) permit the existence of the following motion integrals of the particle number N, the momentum P and the energy E:

$$N = \int_{-\infty}^{\infty} |\psi|^2\,\mathrm{d}x, \tag{1.64}$$

$$P = \int_{-\infty}^{\infty} [\mathrm{i}(\psi\psi_x^* - \psi^*\psi_x) + 2nu]\,\mathrm{d}x, \tag{1.65}$$

$$E = \int_{-\infty}^{\infty} [|\psi_x|^2 + n|\psi|^2 + \tfrac{1}{2}n^2 + \tfrac{1}{2}u^2]\,\mathrm{d}x, \qquad n_t + u_x = 0. \tag{1.66}$$

It can be shown that the energy of the low-frequency system decreases upon the formation of the bound system ψ_s and n_s, i.e. such a state is energetically profitable.

1.8 DIRECT PERTURBATION THEORY METHODS FOR SOLITONS

In this section we will outline a version of the perturbation theory method for perturbed nonlinear wave equations. The peculiarity of this method, to be referred to as a direct one in the following, lies in the fact that it is applicable to the description of perturbed integrable as well as nonintegrable systems.

Below we will consider systems that with zero perturbation turn into nonintegrable ones which have solitary wave solutions. As an example, a perturbed nonlinear Klein–Gordon equation

$$\varphi_{tt} - \varphi_{xx} + u'(\varphi) = \varepsilon R(\varphi) \tag{1.67}$$

can serve.

For $\varepsilon = 0$ and $u'(\varphi) = -\varphi + \varphi^3$ we have a nonintegrable φ^4 model possessing kink ($\sigma = 1$) and antikink ($\sigma = -1$) solutions.

To study the behavior of solutions for these equations we will apply a version of the asymptotic method of small-parameter expansion put forward by Gorshkov and Ostrovsky [1.23]. This method allows the analysis of the evolution of single solitons and soliton complexes under weak perturbations, irrespective of whether the initial unperturbed system is integrable or not.

Here we present the basic ideas of their approach and illustrate its application to a number of typical examples encountered in problems of solitons in inhomogeneous media and dynamic stochasticity in the soliton motion (Chapter 5). Following Ref. [1.23] we will examine a set of differential field equations of the first order in time and space:

$$M(\varphi, \varphi_t, \nabla\varphi; \tau, \boldsymbol{\rho}, \varepsilon) = 0. \tag{1.68}$$

Here φ is a vector function, $\varphi = \{\varphi_1, \ldots, \varphi_N\}$, $\varepsilon \ll 1$, and $\tau = \varepsilon t$ and $\boldsymbol{\rho} = \varepsilon \mathbf{x}$ are slow time and coordinates, respectively.

Further we will study the evolution of steady-state particular solutions of (1.68) which for $\varepsilon = 0$ have the form

$$\varphi = \varphi^{(0)}(\zeta, A), \tag{1.69}$$

where $A = \{A_1 \cdots A_m\}$ are arbitrary constants. In the following we will be interested in solutions for which the following conditions at infinity in ζ

$$\varphi^{(0)}(\zeta, A) = \varphi^{(\pm)(0)}, \qquad \zeta \to \pm\infty,$$

are met. Bearing in mind the main idea of the small-parameter method, we will seek the solution as an asymptotic series,

$$\varphi(x, t) = \varphi^{(0)}(\zeta, A, \boldsymbol{\rho}, \tau) + \sum_{n=1}^{N} \varepsilon^n \varphi^{(n)}(\zeta, \boldsymbol{\rho}, \tau), \tag{1.70}$$

where N is the order of expansion. Note that A and v are functions of slow $\boldsymbol{\rho}$ and τ. Later on, the procedure is effected in a similar manner as in that by Bogolyubov and Krylov in nonlinear oscillation theory, i.e. (1.70) is substituted into the initial (1.68) and the expansion terms are equated at each order in ε to yield a linear set of equations for the corrections $\varphi^{(n)}$:

$$\begin{aligned} &\hat{L}\varphi^{(n)} = H^{(n)}, \\ &\hat{L} = \frac{\partial M^{(0)}}{\partial \varphi_\zeta} \frac{\mathrm{d}}{\mathrm{d}\zeta} + \frac{\partial M^{(0)}}{\partial \varphi}, \\ &M^{(0)} = M(\varepsilon = 0, \varphi = \varphi^{(0)}), \\ &H^{(1)} = -\frac{\partial M}{\partial \varepsilon} - \frac{\partial M}{\partial \varphi_t} \varphi^{(0)}_\tau - \frac{\partial M}{\partial \nabla \varphi} \nabla_\rho \varphi^{(0)}_\rho. \end{aligned} \tag{1.71}$$

It is inferred from (1.71) that the solution for $\varphi^{(n)}$ has the form

$$\varphi^{(n)} = Y\left(C^{(n)} + \int_0^\zeta \mathrm{d}\zeta' Y^+ H^{(n)} \right), \tag{1.72}$$

where $C^{(n)} = \text{constant}$, Y is a matrix for the fundamental system of the solutions of the equation

$$\hat{L}Y = 0.$$

and

$$Y^{+} = Y^{-1}\left[\frac{\partial M^{(0)}}{\partial \varphi_{\zeta}}\right]^{-1}. \tag{1.73}$$

Note that the Y_i are found from the known $\varphi^{(0)}$ by varying, with respect to ξ and A, the generated system

$$Y_1 = \varphi^{(0)}_{\zeta}, \qquad Y_i = \varphi^{(0)}_{A_i}, \quad i = 1, 2, 3, \ldots, m + 1. \tag{1.74}$$

For the expansion (1.70) to exist one has to impose the conditions of boundedness on the terms of this series that are derived from (1.72). Among the terms of (1.72) one can see exponentially growing Y_i. Hence, the finiteness of the solutions for $\varphi^{(n)}$ can be achieved in two ways:

(a) For an odd function $Y^{+}H^{(n)}$, the solution for $\varphi^{(n)}$ is finite if the constants $C^{(n)}$ satisfy the relation

$$C^{(n)}_{\alpha} = \int_0^{\infty} \mathrm{d}\zeta\, Y^{+}_{\alpha} H^{(n)}. \tag{1.75}$$

(b) For an even function $Y^{+}H^{(n)}$, the condition

$$\int_{-\infty}^{\infty} \mathrm{d}\zeta\, Y^{+}_{\alpha} H^{(n)} = 0 \tag{1.76}$$

must be satisfied. A secular growth of $\varphi^{(n)}$, due to finite solutions of (1.74) for $\zeta \to \pm\infty$, is ruled out by the conditions

$$\lim Y^{+}_{i} H^{(n)} = 0, \qquad i = l + 1, l + 2, \ldots, m + 1, \quad \zeta \to \pm\infty. \tag{1.77}$$

Let us illustrate the application of this method. Consider a system described by the nonlinear Klein–Gordon equation

$$\varphi_{tt} - \varphi_{xx} - F(\varphi) = \varepsilon R(\varphi). \tag{1.78}$$

Here $F(\varphi)$ is an arbitrary nonlinear function of φ having at least a pair of zeros; $R(\cdot)$ is a nonlinear operator, and $R(\varphi^{(0)}) \to 0$ for $\zeta \to \pm\infty$. The soliton solution of (1.78) for $\varepsilon = 0$ has the form

$$\varphi^{(0)} = \varphi[(1 - v^2)^{-1/2}\zeta], \qquad \zeta = x - vt. \tag{1.79}$$

With the aid of the expansion (1.70), the linear operator L is found to be

$$\hat{L} = (v^2 - 1)\frac{\mathrm{d}^2}{\mathrm{d}\zeta^2} - F'(\varphi^{(0)}). \tag{1.80}$$

There is a solution for $\hat{L}y = 0$ that decreases for

$$\zeta \to \pm\infty, \qquad y_1 = \varphi^{(0)}(\zeta).$$

Hence we deduce that the solution each approximation can be written as follows:

$$\varphi^{(n)} = y_1\left(C_1^{(n)} + \int_0^\zeta d\zeta' y_2 H^{(n)}\right) + y_2\left(C_2^{(n)} + \int_0^\zeta d\zeta' y_2 H^{(n)}\right),$$

where

$$y_2 = \varphi_\zeta^{(0)} \int_0^\zeta d\zeta' [\varphi_\zeta^{(0)}]^{-2}. \tag{1.81}$$

For $\varphi^{(n)}$ to be finite it is necessary and sufficient that the orthogonality condition,

$$\int_{-\infty}^{\infty} d\zeta \varphi_\zeta^{(0)} H^{(n)} = 0, \tag{1.82}$$

be met. For $n = 1$

$$H^{(1)} = \varphi_\zeta^{(0)} \frac{dv}{dt} + 2v\varphi_{\zeta\tau}^{(0)} + \varepsilon R(\varphi^{(0)}).$$

Substitution of this expression into (1.82) results in the equation for soliton velocity v:

$$\frac{d}{dt}\left(\frac{v}{\sqrt{(1-v^2)}}\right) = -\frac{\varepsilon}{\langle \varphi_z^2 \rangle} \int dz \varphi_z R(\varphi^{(0)}), \tag{1.83}$$

where

$$Z = \frac{\zeta}{\sqrt{(1-v^2)}}, \qquad \langle \varphi_z^2 \rangle = \int_{-\infty}^{\infty} dz \varphi_z^2.$$

In a similar way one can derive equations describing the soliton evolution in the Zakharov perturbed system, generalized KdV and many other systems. These cases will be described in the next sections.

A number of authors developed the ideas close to the approach put forward by Gorshkov and Ostrovsky [1.23]. Thus Kodama and Ablowitz [1.24], relying on similar ideas, studied soliton stability of the highly nonlinear KdV equation against various perturbations.

It should be noted that direct methods enjoy the advantage that they are applicable also to cases when the initial equation is nonintegrable. Direct methods also have drawbacks. In particular, it is difficult with them to analyze nonsoliton initial conditions or evaluate high-order corrections in ε, or deal with as the evolution of multisoliton complexes. In the latter cases it seems to be more appropriate to apply perturbation theory methods based on the IST (of course, only if, for $\varepsilon = 0$, the initial system belongs to a class of integrable systems).

Let us apply them to Zakharov's system with nonuniform parameters:

$$n^{(0)} = \tilde{n} - \frac{\varphi^{(0)2}}{1 - v^2}. \tag{1.84}$$

Let $\tilde{n} = \tilde{n}(x, t)$. Then the solution of (1.59)–(1.60) can be sought in the form of a series

$$\psi(x, t) = \left[\varphi^{(0)}(\zeta, \rho, \tau) + \sum_{n=1} \varepsilon^n \varphi^{(n)}(\zeta, \rho, \tau) \right] \exp\left[\mathrm{i}\left(\frac{v}{2}\zeta + \varphi \right) \right],$$

$$n = n^{(0)}(\zeta, \rho; \tau) + \sum_{n=1} \varepsilon^n n^{(n)}(\zeta, \rho, \tau). \tag{1.85}$$

Here $\varphi^{(0)}$, $n^{(0)}$ are soliton solutions; $v = v(t)$ are slow time and space functions, $\rho = \varepsilon x$, $\tau = \varepsilon t$, and ε is a small parameter which is proportional to the ratio between the soliton scale and the given low-frequency wave scale. Substituting (1.85) into (1.73) and (1.59)–(1.60) and equating coefficients with the same powers of ε, we obtain a linear set of equations:

$$\hat{L}_1 \operatorname{Re} \varphi^{(n)} = \left[\frac{\mathrm{d}^2}{\mathrm{d}\zeta^2} - (\lambda^2 + 3n^{(0)}) \right] \operatorname{Re} \varphi^{(n)} = \operatorname{Re} H^{(n)},$$

$$\hat{L}_2 \operatorname{Im} \varphi^{(n)} = \left[\frac{\mathrm{d}^2}{\mathrm{d}\zeta^2} - (\lambda^2 + n^{(0)}) \right] \operatorname{Im} \varphi^{(n)} = \operatorname{Im} H^{(n)}. \tag{186}$$

As Gorshkov and Ostrovsky [1.23] showed, a necessary requirement for the corrections not to grow is orthogonality of $H^{(n)}$ to the eigenfunctions (decreasing to zero for $\zeta \to \pm\infty$) of the L_1 and L_2 adjoint problem. These requirements are written as follows:

$$\langle \varphi_\zeta^{(0)} \operatorname{Re} H^{(n)} \rangle = 0, \qquad \langle \varphi^{(0)} \operatorname{Im} H^{(n)} \rangle = 0, \tag{1.87}$$

where

$$\langle \cdot \rangle \equiv \int_0^t (\cdot)\, \mathrm{d}\zeta.$$

We obtain from this, for soliton parameters, the following:

$$\frac{\mathrm{d}}{\mathrm{d}t}\lambda(1-v^2)=0;$$

$$\lambda(1-v^2)\left(\frac{\mathrm{d}v}{\mathrm{d}t}+2\frac{\partial\tilde{n}}{\partial x}\right)+\frac{8}{3}\frac{\mathrm{d}}{\mathrm{d}t}(v\lambda^3)=0.$$

This system may be rewritten as the second-order equation

$$\frac{\mathrm{d}^2x}{\mathrm{d}t^2}=-2(\tilde{n}_x)\Bigg/\left[1+\tfrac{8}{3}c^2\frac{1+5\dot{x}^2}{(1-\dot{x}^2)^4}\right],$$

where $c=\text{constant}$ is the so-called 'quanta number'.

1.9 IST-BASED PERTURBATION THEORY FOR SOLITONS

If a system refers to a class of nearly integrable systems, i.e. it becomes integrable upon turning off a small parameter, then to investigate soliton dynamics in media (for $\varepsilon \ll 1$) it is often convenient to apply perturbation theory based on the IST. Let us formulate the scheme of considerations for soliton and soliton complexes. Let the problem considered be described by a nearly integrable nonlinear wave equation

$$\hat{N}Lu(x,t)=\varepsilon\hat{R}(u(x,t)), \tag{1.88}$$

where $NL(u(x,t))$ is an operator corresponding to an integrable nonlinear wave equation. In specific cases it has the following forms:

for the KdV equation,

$$\hat{N}L=\partial_t\pm 6u\partial_x+\partial_x^3,$$

for the SG equation,

$$\hat{N}L=\partial_t^2-\partial_x^2+\sin(\cdot),$$

for the nonlinear string equation,

$$\hat{N}L=\partial_t^2-\partial_x^2+\partial_x^4+(u_x^2)_x\partial_x,$$

and so on. The operator $R(u)$ stands for perturbations. For initial conditions as solitons and their complexes (breathers, bions and others) based on the IST (for $\varepsilon \neq 0$), one can derive a set of equations for their parameters (amplitudes, velocities, frequencies, relative separations between solitons, etc). This is a coupled system of nonlinear ordinary differential equations that is finite-dimensional. The latter fact allows the application of well-developed methods for the analysis of finite-dimensional dynamic systems.

Here we outline briefly the main schemes of considerations which allow the derivation of equations for the parameters of solitons and soliton complexes in nearly integrable systems, referring for details to Refs. [1.25]–[1.28].

Let us consider the evolution of a single soliton of the perturbed sine–Gordon equation [1.28]

$$u_{tt} - u_{xx} + \sin u = \varepsilon f(x,t)R(u), \qquad \varepsilon \ll 1. \tag{1.89}$$

The use of the IST relies on the representation of (1.89) for $\varepsilon = 0$ as an L–A pair. The operators $\hat{L}$and $\hat{A}$ are

$$\hat{L} = \frac{\partial}{\partial x} - \frac{\mathrm{i}}{2}\left[\left(\lambda - \frac{1}{4\lambda}\cos u\right)\sigma_3 - \frac{\sigma_2}{4\lambda}\sin u + \frac{\sigma_1}{2}(u_x - u_t)\right], \tag{1.90}$$

$$\hat{A} = \frac{\partial}{\partial t} + \frac{\mathrm{i}}{2}\left[\left(\lambda + \frac{1}{4\lambda}\cos u\right)\sigma_3 + \frac{\sigma_2}{4\lambda}\sin u + \frac{\sigma_1}{2}(u_x - u_t)\right]. \tag{1.91}$$

Here the σ_i are Pauli matrices, λ is a spectral parameter. The Jost coefficients $a(\lambda)$, $b(\lambda)$ and the discrete spectrum λ_n determine scattering data. Their time dependence is defined by the expressions

$$\begin{aligned} &a(\lambda,t) = a(\lambda,0), \qquad b(\lambda,t) = b(\lambda,0)\exp\left[-2\mathrm{i}\omega(\lambda)t\right] \\ &\lambda_n(t) = \lambda_n(0), \qquad b_n(\lambda,t) = b_n(0)\exp\left[-2\mathrm{i}\omega(\lambda_n)t\right] \quad \omega(\lambda) = \tfrac{1}{2}(\lambda + \tfrac{1}{4}\lambda). \end{aligned} \tag{1.92}$$

From these data the potential $u(x,t)$ is recovered. For $\varepsilon \neq 0$ one can develop a perturbation theory relying on the correspondence between (1.89) and a direct scattering problem. But in this case the time dependence of scattering data is essentially modified. Using the expression for the variational derivatives of scattering data, one can obtain a set of equations [1.1], [1.28]:

$$\frac{\mathrm{d}a(\lambda,t)}{\mathrm{d}t} = -\frac{\mathrm{i}\varepsilon}{4}\int_{-\infty}^{\infty} \mathrm{d}x R(u) W(\psi,\sigma,\varphi) \tag{1.93}$$

$$\frac{\mathrm{d}b(\lambda,t)}{\mathrm{d}t} = -2\mathrm{i}\omega(\lambda)b(\lambda,t) - \frac{\mathrm{i}\varepsilon}{4}\int_{-\infty}^{\infty} \mathrm{d}x R(u) W(\sigma,\varphi,\tilde{\psi}) \tag{1.94}$$

$$\frac{\mathrm{d}\lambda_n}{\mathrm{d}t} = \frac{\mathrm{i}\varepsilon}{4a'(\lambda_n)} \times \int_{-\infty}^{\infty} \mathrm{d}x R(u) W[\psi(\lambda_n), \sigma_1\varphi(\lambda_n)], \tag{1.95}$$

$$\frac{\mathrm{d}b_n}{\mathrm{d}t} = -2\mathrm{i}\omega(\lambda_n)b_n - \frac{\mathrm{i}\varepsilon}{4a'(\lambda_n)}\int_{-\infty}^{\infty} \mathrm{d}x R(u) W(\sigma_1\varphi(\lambda_n), \varphi'_n - b_n\psi'_n). \tag{1.96}$$

Here $W(\varphi,\psi)$ is the Wronskian ψ and φ are the Jost functions. In the case of a single soliton, $\sigma = \pm 1$ (kink or antikink)

$$u_{\mathrm{s}}(z) = 4\tan^{-1}(\exp z),$$

$$z = \frac{x-\zeta}{\sqrt{(1-v^2)}}, \qquad \nu = \tfrac{1}{2}\sqrt{\left(\frac{1+v}{1-v}\right)},$$

$$\lambda = \mathrm{i}\nu, \qquad c_1 = \frac{b_1}{a'(\mathrm{i}\nu)} = -2\sigma\nu\exp\left(\frac{\zeta}{\sqrt{(1-v^2)}}\right), \quad a' = \frac{\partial a}{\partial\lambda}.$$

The Jost functions have the forms

$$\psi_{\mathrm{s}} = \mathrm{e}^{\mathrm{i}k(\lambda)x}\begin{pmatrix}\lambda + \mathrm{i}\nu\tanh z\\ \sigma\nu\,\mathrm{sech}\, z\end{pmatrix}(\lambda+\mathrm{i}\nu)^{-1},$$

$$\varphi_{\mathrm{s}} = \mathrm{e}^{-\mathrm{i}k(\lambda)x}\begin{pmatrix}-\sigma\nu\,\mathrm{sech}\, z\\ \lambda - \mathrm{i}\nu\tanh z\end{pmatrix}(\lambda+\mathrm{i}\nu)^{-1},$$

$$a_{\mathrm{s}}(\lambda) = \frac{\lambda - \mathrm{i}\nu}{\lambda+\mathrm{i}\nu}, \qquad b_{\mathrm{s}}(\lambda) = 0. \tag{1.97}$$

From this, within the adiabatic approximation, one can derive an equation for the soliton parameters, namely

$$\frac{dv}{dt} = -\frac{\varepsilon\sigma}{4}(1 - v^2)^{3/2} \int_{-\infty}^{\infty} dz\, R(u_s(z)) \operatorname{sech} z \tag{1.98}$$

$$\frac{d\zeta}{dt} = v - \frac{v\varepsilon\sigma(1 - v^2)}{4} \int_{-\infty}^{\infty} dz\, zR(u_s(z)) \operatorname{sech} z. \tag{1.99}$$

For $a(\lambda)$, $b(\lambda)$ see Appendix 2.

This scheme permits the determination of the contribution at the expense of higher orders over ε, namely, ignoring the contribution of the continuous spectrum to (1.98) and (1.99).

Should perturbation frequencies Ω_i be close to $\omega_0 = 1$, the excitation of linear waves in the system is probable. In this case the equations in the adiabatic approximation are to be modified [1.29], [1.30] (see Appendix 2).

As another example, we will present perturbation theory formulae for nearly integrable modifications of the Landau–Lifshitz equation [1.31]:

$$\frac{\partial \mathbf{S}}{\partial t} = [\mathbf{S} \times \mathbf{S}_{xx}] + [\mathbf{S} \times J\mathbf{S}] + \varepsilon f(x, t)\mathbf{R}(\mathbf{S}), \tag{1.100}$$

where a diagonal matrix $J = \operatorname{diag}(J_1, J_2, J_3)$ has been introduced. The case $J_1 = J_2 = J_3$ corresponds to an isotropic ferromagnet, $J_1 = J_2$ refers to a ferromagnet having 'easy axis' anisotropy, and $J_2 = J_3$ pertains to a ferromagnet with 'easy plane' anisotropy. Later on we put

$$\beta_1 = J_1 - J_2, \qquad \beta_3 = J_3 - J_2, \qquad J_1 \leqslant J_2 \leqslant J_3.$$

For conservation of the modulus of the vector $S(x, t)$, we require that $(\mathbf{S}, \mathbf{R}(\mathbf{S})) = 0$. As zero approximation, $\varepsilon = 0$, $\mathbf{S}^{(0)} = (S_1^{(0)}, S_2^{(0)}, S_3^{(0)})$, we will choose the one-soliton solution of the unperturbed equation (1.100),

$$\begin{aligned} &S_1^{(0)} + iS_2^{(0)} = e^{i\varphi} \operatorname{sech} z, \qquad && S_3^{(0)} = \kappa \tanh z, \\ &Z = \xi x - \kappa\tau t + \text{const.}, && \varphi = \text{const.} \end{aligned} \tag{1.101}$$

Here κ is the parameter characterizing the polarity of the soliton. For solitons $\kappa = 1$, for antisolitons $\kappa = -1$. The angle φ describes the orientation of the vector $S^{(0)}$ in the plane perpendicular to the anisotropy axis (0, 0, 1). Parameter ξ and τ depend on J_α and φ through the relations

$$\begin{aligned} \xi &= \xi(\varphi) = \sqrt{(\beta_3 + |\beta_1| \cos^2 \varphi)}, \\ \tau &= \tau(\varphi) = -|\beta_1| \sin \varphi \cos \varphi. \end{aligned} \tag{1.102}$$

Let us proceed now with new variables $z = \xi(x - X)$. In this case the equations of the adiabatic approximation have the form

$$\frac{d\varphi}{dt} = -\frac{\varepsilon}{2}\int_{-\infty}^{\infty} dz \frac{R_-(z)}{\cosh z}, \tag{1.103}$$

$$\frac{dX}{dt} = \frac{\kappa\tau}{\xi} - \frac{\varepsilon\tau}{2\xi^3}\int_{-\infty}^{\infty} dz \frac{zR_-(z)}{\cosh z} + \frac{\varepsilon}{2\xi}\int_{-\infty}^{\infty} dz \frac{R_+(z)}{\sinh z}. \tag{1.104}$$

Here

$$R_{\mp}(z) = \begin{pmatrix} \sin\varphi \\ \cos\varphi \end{pmatrix} R_1(\mathbf{S}^{(0)}) \mp \begin{pmatrix} \cos\varphi \\ \sin\varphi \end{pmatrix} R_2(\mathbf{S}^{(0)}). \tag{1.105}$$

Equation (1.104) defines the domain wall energy, changing under perturbation (1.100). It results from the equation

$$\frac{dH_0}{dt} = \varepsilon \int_{-\infty}^{\infty} dx \left(R(\mathbf{S}), \left[\mathbf{S} \times \frac{\partial \mathbf{S}}{\partial x} \right] \right),$$

where H_0 is a unperturbed Hamiltonian. Substituting in this equation the adiabatic form of the solution (1.101), we obtain equation (1.104). In a similar manner one can study the multidimensional SG equation [1.32], the perturbed NLSE and other wave equations. A summary of the basic perturbation theory formulae is given in Appendices 1–3.

1.10 ASYMPTOTIC METHODS FOR NONLINEAR PERIODIC WAVES

Previously we studied perturbation theory methods, which are applicable to the evolution of soliton and breathers under perturbations. In this case, the evolution of nonlinear periodic waves (NPW), these methods ought to be modified. Nowadays two types of methods have been developed. The small-parameters method is typical of the first group (see Section 1.8). Its application to the NPW dynamics under perturbations is outlined in Ref. [1.33], where a system of equations for the wave parameters is derived.

Similar equations result also by the Whitham averaging method [1.35], relying on the representation of an averaged Lagrangian of the nonlinear wave field.

Let us describe briefly the application of direct methods to the evolution of nonlinear periodic waves under perturbations [1.33].

Consider a set of equations

$$\frac{\partial}{\partial t}\frac{\partial L}{\partial u_t}+\nabla\frac{\partial L}{\partial \nabla u}-\frac{\partial L}{\partial u}=\Phi, \qquad \Phi=\Phi^{(0)}+\varepsilon\Phi^{(1)}+O(\varepsilon^2). \tag{1.106}$$

Here $u(x,t)$ is an N-dimensional vector function, L is the Lagrangian, Φ is force density, $\varepsilon \ll 1$. It is assumed that L and Φ are functions of slow time $\tau=\varepsilon t$ and space coordinate $\boldsymbol{\rho}=\varepsilon\mathbf{x}$.

Let (1.106) for $\Phi=\Phi^{(0)}$ have solutions as plane waves, $u=U(\theta)$, $\theta=\omega t-\mathbf{kx}+\theta_0$, which result from a set of equations

$$\frac{\partial}{\partial t}\frac{\partial L}{\partial u_t}+\nabla\frac{\partial L}{\partial \nabla u}-\frac{\partial L}{\partial u}=\Phi^{(0)}(\tau,\rho=\text{const.}), \tag{1.107}$$

and $U(\theta)$ depend on integration constants θ_0 and A and on the parameters $\tau,\boldsymbol{\rho},\omega$ and $\mathbf{k}$. Further,

$$\omega=\omega(\mathbf{k},A).$$

U is a 2π-periodic function of θ. A solution will be sought as a function close to $U(\theta)$,

$$u=U(\theta,A_i,\tau,\boldsymbol{\rho})+\sum_{n=1}^{\infty}\varepsilon^n u^{(n)}(\theta,\tau,\boldsymbol{\rho}),$$

$$A(\tau,\boldsymbol{\rho})=\sum_{n=0}^{\infty}\varepsilon^n A^{(n)}(\tau,\boldsymbol{\rho}),$$

$$\theta=\theta^{(0)}(t,\mathbf{r},\tau,\boldsymbol{\rho})+\sum_n \varepsilon^n\theta^{(n)}(\tau,\boldsymbol{\rho}). \tag{1.108}$$

Then expanding L and Φ in series in ε, and taking (1.108) into account, one can derive from the motion equations (1.106) and (1.107), by equating the coefficients with the same power of ε, a linear set of equations to define the functions A, θ and $U^{(n)}$ within the small ε approximation. It has the form

$$T\left(\tau,\boldsymbol{\rho},\theta,\frac{\partial}{\partial\theta},\frac{\partial^2}{\partial\theta^2}\right)u^{(n)}=H^{(n)}(\tau,\rho,\theta), \tag{1.109}$$

$$\begin{aligned}Tu^{(n)}=&\frac{\partial}{\partial\theta}\left[\frac{\partial^2 L}{\partial u_\theta^2}u_\theta^{(n)}+\frac{\partial^2 L}{\partial u_\theta\partial u}u^{(n)}\right]\\&-\frac{\partial^2 L}{\partial u^2}u^{(n)}-\frac{\partial^2 L}{\partial u\partial u_\theta}u_\theta^{(n)}-\frac{\partial\Phi^{(0)}}{\partial u}u^{(n)}-\frac{\partial\Phi^{(0)}}{\partial u_\theta}u_\theta^{(n)},\end{aligned} \tag{1.110}$$

$$H^{(1)} = \phi^{(1)} + \frac{\partial \Phi^{(0)}}{\partial u_t} u_\tau + \frac{\partial \Phi^{(0)}}{\partial \nabla u} \nabla_\rho u + \frac{\partial^2 L}{\partial u \partial u_t} u_\tau + \frac{\partial^2 L}{\partial u \partial \nabla u} \nabla_\rho u - \frac{\partial}{\partial \tau} \frac{\partial L}{\partial u_t}$$
$$- \nabla_\rho \frac{\partial L}{\partial \nabla u} - \omega \frac{\partial}{\partial \theta} \left[\frac{\partial^2 L}{\partial u_t^2} u_\tau + \frac{\partial^2 L}{\partial u_t \partial \nabla u} \right]$$
$$+ \bar{k} \frac{\partial}{\partial \theta} \left[\frac{\partial^2 L}{\partial \nabla u \partial u_t} u_\tau + \frac{\partial^2 L}{\partial (\nabla u)^2} \nabla_\rho u \right].$$

The solution of (1.109) is

$$u^{(n)} = Y C^{(n)} + Y \int_0^\theta Y^* H^{(n)} \, \mathrm{d}\theta'. \tag{1.111}$$

Here $C^{(n)}$ is a constant vector, Y is a matrix composed of the vectors of the fundamental system of solutions of the variational equations $T\psi = 0$, and Y^* is a matrix for the adjoint system

$$\frac{\mathrm{d}Y}{\mathrm{d}\theta} Y^* = \left[\frac{\partial^2 L}{\partial u_\theta^2} \right]^{-1}, \qquad Y Y^* = 0.$$

After some modifications, one can show that $U^{(n)}$ is bonded by the function θ, provided the following conditions hold:

$$C_j^{(n)} = \begin{cases} -\displaystyle\int_0^\infty y_{jk}^* H_k^{(n)} \mathrm{e}^{-\lambda_j \theta} \, \mathrm{d}\theta, & \operatorname{Re} \lambda_j > 0 \\ \displaystyle\int_{-\infty}^0 y_{jk}^* H_k^{(n)} \mathrm{e}^{-\lambda_j \theta} \, \mathrm{d}\theta, & \operatorname{Re} \lambda_j < 0. \end{cases} \tag{1.112}$$

$$2\pi\alpha C_2^{(n)} + \int_0^{2\pi} y_{1k}^* H_k^{(n)} \, \mathrm{d}\theta + \alpha \int_0^{2\pi} \mathrm{d}\theta \int_0^\theta y_{2k}^* H_k^{(n)} \, \mathrm{d}\theta' = 0, \tag{1.113}$$

$$\int_0^{2\pi} y_{2k}^* H_k^{(n)} \, \mathrm{d}\theta = 0. \tag{1.114}$$

It is convenient also to use the equation

$$\int_0^{2\pi} y_{1k}^* H_k^{(n)} \, \mathrm{d}\theta = 0, \tag{1.115}$$

allowing the possibility of performing the transition to a quasiharmonic wave. Equations (1.111)–(1.115) are useful for the calculation of $A^{(n)}, C_j^{(n)}, u^{(n)}$. However, unlike the case of solitons and breathers direct methods turned out to be inefficient for the analysis of chaotic phenomena in NPW dynamics. They are applicable to the extreme cases when an NPW may be treated as a chain of weakly corrected solitons. Relevant applications will be studied in Chapter 5.

So far, basic results on dynamic NPW stochastization have been obtained by the Hamiltonian approach. This approach is comprehensively outlined in a book by Zaslavsky [1.34], to which the reader is referred for details. Here we will only briefly describe it. The initial point here is the representation of a linear wave as a wave packet composed of strongly localized plane waves. For weak perturbations, the NPW evolution can be reduced to the analysis of the parameter evolution of each separate harmonic. This investigation will be carried out on the basis of the Hamiltonian approach. As an example we will consider a perturbed φ^4 model, following a method put forward in Ref. [1.34].

The initial wave has the form

$$\varphi_{tt} - \varphi_{xx} + \varphi^3 - \varphi = \varepsilon F(\varphi(x,t)), \qquad \varepsilon \ll 1. \tag{1.116}$$

For $\varepsilon = 0$, nonlinear periodic solutions to this equation are (1.52):

$$\varphi_0(x,t) = a\,\mathrm{sn}\left[a\zeta/\sqrt{[2(1-v^2)]}, \frac{b}{a}\right]$$

$$a = 1 + \sqrt{(1+2c)}, \qquad b = 1 - \sqrt{(1-2c)}. \tag{1.117}$$

The wave period is $4K(k)$, $k = b/a$. This solution is expanded in a Fourier series, with the form

$$\varphi_0(x - vt) = \sum_{n=-\infty}^{\infty} a_n \exp[ink(x - vt)]. \tag{1.118}$$

Here

$$a_n = \frac{q^{n-1/2}}{1 - q^{2n-1}} \frac{2\pi}{kK(k)}; \qquad q = \mathrm{e}^{-\pi k'(k)/K(k)}, \tag{1.119}$$

where $K(k)$ is the first-order elliptic integral, and a_n is seen to decrease rapidly with the growth of n. We can determine a typical number of harmonics in the wave spectrum, N. In our case

$$N \simeq \tfrac{1}{2} \ln \frac{16}{1 - k^2}. \tag{1.120}$$

To define the variation of harmonic parameters under perturbation, $\varphi(x,t)$ and $F(\varphi, x, t)$ are expanded in Fourier series:

$$\varphi(x,t) = \sum_n \varphi_n \exp(ink_0 x), \qquad k_0 = \frac{2\pi}{\lambda}. \tag{1.121}$$

$$F(\varphi) = \sum_n F_n \exp(ink_0 x),$$

substitution of (1.121) into (1.116) leads to an infinite set of equations:

$$\ddot{\varphi}_n + k^2 n^2 \varphi_n - \varphi_n + \sum_{n_1, n_2, n_3} \varphi_{n_1}\varphi_{n_2}\varphi_{n_3}\delta(n - n_1 - n_2 - n_3) = \varepsilon F_n. \quad (1.122)$$

These can be written in the Hamiltonian form

$$\frac{\mathrm{d}\dot{\varphi}_n}{\mathrm{d}t} = -\frac{\partial H}{\partial \varphi_{-n}}, \qquad \frac{\mathrm{d}\varphi_n}{\mathrm{d}t} = \frac{\partial H}{\partial \dot{\varphi}_{-n}}, \quad (1.123)$$

where the Hamiltonian has the form $H = H_0 + \varepsilon H_1$, with

$$H_0 = \frac{1}{2}\sum_n (\dot{\varphi}_n\dot{\varphi}_{-n} + n^2 k_0^2 \varphi_n \varphi_{-n}) + \left(-\frac{1}{k}\right)\sum_{n_1, n_2} \varphi_{n_1}\varphi_{n_2}\delta(n_1 + n_2) \quad (1.124)$$

$$+ \frac{1}{k}\sum \varphi_{n_1}\varphi_{n_2}\varphi_{n_3}\varphi_{n_4}\delta(n_1 + n_2 + n_3 + n_4);$$

$$\varepsilon H_1 = \varepsilon \sum_n \sum_{k=0}^{\infty} \psi_n^{(k)}(t) \sum_{n_1, \cdots n_k} \varphi_{n_1} \cdots \varphi_{n_k}\delta(n_1 + \cdots + n_k), \quad \frac{\partial \psi}{\partial \varphi} = F.$$

Let us consider NPW solutions as an unperturbed Hamiltonian. The relevant Hamiltonian is H_0, and $\varphi(x, t)$ is expanded in the Fourier series taking into account that it is a function of $(x - vt)$,

$$\varphi(x, t) = \sum_n a_n \exp[in(kx - \theta)], \qquad \varphi_n = a_n \mathrm{e}^{-in\theta}.$$

In the case when the perturbation F is independent of φ, the evolution of the nonlinear wave can be shown to be described by reduced equations

$$\frac{\mathrm{d}H_0}{\mathrm{d}t} = -\mathrm{i}\varepsilon\omega(H_0)\sum n a_n(H_0)\mathrm{e}^{in\theta}F_n(t),$$

$$\frac{\mathrm{d}\theta}{\mathrm{d}t} = \omega(H_0). \quad (1.125)$$

Equations (1.125) allow the analysis of nonlinear resonances between the wave and the external perturbation, the determination of the NPW parameter modulation, and the determination of the criteria for stochastic wave instability.

2

ADIABATIC DYNAMICS OF SOLITONS

2.1 THE MOTION OF SINE–GORDON SOLITONS IN AN INHOMOGENEOUS MEDIUM

Many authors have pointed out an analogy between solitons and particles. First it was discovered by Zabusky and Kruskal in their famous numerical experiments [2.1], where they observed the elastic collision between the KdV solitons. This circumstance in particular suggested the choice of the notion 'soliton' for these localized wave packets. Since that time a number of effects in which solitons behave like particles have been found (in a periodic field, under the action of damping, etc. [2.2]).

The soliton motion in an inhomogeneous medium is another example where the soliton exhibits properties of a classical particle. Reference [2.2] was one of the first works in which soliton propagation in an inhomogeneous medium has been studied and SG solitons under the action of various perturbations have been considered. Here we consider the problem of SG soliton propagation in a medium with varying parameters [2.3]. The corresponding wave equation has the form

$$u_{tt} - u_{xx} + (1 + \varepsilon \cos ax) \sin u + \Gamma u_t = f_0 \sin \omega t. \tag{2.1}$$

Here $u(x, t)$ is the dimensionless magnetic flux, normalized to the value of the quantum of magnetic flux $\Phi_0 = hc/2e$, a is the period of the inhomogeneity, $f_0 = I_a/I_c$ is the external field, Γ is the dissipation coefficient. This equation describes the motion of the Josephson vortex in a periodic inhomogeneous long junction under the action of a high-frequency external force, with damping taken into account as well as solitons in inhomogeneous magnetic media, liquid crystals, etc. The junction inhomogeneity along the direction of the vortex motion can be produced by the periodic modulation of transition, and the external force by the alternating current applied perpendicular to the junction. The junction is supposed to be long, i.e. $L \gg \lambda_J$, where L is the junction length, λ_J is the Josephson penetration depth. The penetration depth of the Josephson junction is taken to be the length scale, the inverse Josephson plasma frequency ω_J to be the time scale:

$$t = t'\omega_J, \qquad x = x'/\lambda_J.$$

The fluxon propagation velocity is divided by the propagation velocity of the electromagnetic wave in the junction, which is $c_0 = \lambda_J/\omega_J$. Also I_c is the critical Josephson current and I_a is the amplitude of the inhomogeneous external alternating current. The notation is thus introduced. At $\varepsilon = \Gamma = f_0 = 0$, the equation (2.1) takes the solitonic (kink) solution (1.26):

$$u = 4\tan^{-1}\left\{\exp\left(\pm\frac{x - vt - \zeta_0}{\sqrt{(1 - v^2)}}\right)\right\}, \tag{2.2}$$

The smallness of the parameters ε, Γ, $f_0 \ll 1$ allows us to consider the dynamics of a single soliton in terms of IST-based perturbation theory (see Section 1.9). Note that its direct application is rather cumbersome because of perturbation delocalization. It is necessary to isolate the induced field caused by the external field [2.4, 2.5]. It takes the form

$$u(x, t) = U(x, t) + \varphi(t). \tag{2.3}$$

Substituting (2.1.3) into (2.1.1) one finds a set of equations

$$U_{tt} - U_{xx} + \sin U = (1 - \cos U)\varphi + \Gamma U_t + \varepsilon\cos(ax)\sin U, \tag{2.4}$$

$$\varphi_{tt} + \varphi = f_0 \sin\Omega t. \tag{2.5}$$

The solution of the linearized equation (2.5) has the form

$$\varphi(t) = \frac{\sin\Omega t}{1 - \Omega^2}. \tag{2.6}$$

Then the equation for $U(t)$ takes the form

$$U_{tt} - U_{xx} + \sin U = f_0(1 - \cos U)\sin(\Omega t)/(1 - \Omega^2) - \varepsilon\cos(ax)\sin U - \Gamma u_t. \tag{2.7}$$

Without violating the basic character of the equation we assume $\Omega^2 \ll 1$ in equation (2.7) which describes a slowly varying field, so that we can apply the perturbation theory developed in Chapter 1. As a result, we obtain equations defining the parameter dependence of v and ζ upon time:

$$\frac{\mathrm{d}v}{\mathrm{d}t} = -(1-v^2)^{3/2}\left[\frac{\Gamma v}{\sqrt{(1-v^2)}} + \frac{\pi\varepsilon\omega^2 \sin(a\zeta)}{4\sinh(\pi\omega/2)}\right] - \frac{\pi f_0 \sin\Omega t}{8}, \tag{2.8}$$

$$\frac{\mathrm{d}\zeta}{\mathrm{d}t} = v + \pi v(1-v^2)\varepsilon\omega\frac{4-\pi\omega\coth(\pi\omega/2)}{8\sinh(\pi\omega/2)}\cos(a\zeta), \qquad \omega = a\sqrt{(1-v^2)}, \tag{2.9}$$

where $f_0 < \varepsilon$, Γ. Then, let us assume that the kink velocity is small, i.e. $v^2 \ll 1$. As will be shown below (Chapter 3), at small kink velocities and under the condition that the wave intensity radiated by the kink is exponentially small during the motion in the inhomogeneous medium, one can make use of the adiabatic approximation. The set (2.8), (2.9) is rewritten in the form

$$\frac{\mathrm{d}v}{\mathrm{d}t} = -\Gamma v + \frac{\pi\varepsilon a^2 \sin a\zeta}{4\sinh(\pi a/2)} - \frac{\pi}{8}f\sin\Omega t, \tag{2.10}$$

$$\frac{\mathrm{d}\zeta}{\mathrm{d}t} = v. \tag{2.11}$$

This set of equations coincides with the equation describing dynamics of a mathematical pendulum under the action of an oscillating external force and damping. This problem has been studied quite well (see, for example, [2.6]). For small velocities the kink length $l_1 \sim 1$. We infer from (2.11) that the depth of the potential well depends strongly on the ratio between the kink length and the modulation medium period a. If $l_k \gg a$, we have an exponentially small effective potential, i.e. the soliton does not feel the inhomogeneity. Here some regimes corresponding to oscillations and rotations can exist. For the soliton they correspond to trapping of solitons by the inhomogeneity. In the other case, $l_k \ll a$, solitons perform unbounded motions. In accordance with the theory of nonlinear resonances, regimes are possible in which the soliton oscillation in the field of the inhomogeneity resonates with the variable field and performs a more complicated motion. Let us analyze this more complicated dynamics on the basis of singular perturbation theory [2.5], given in

Chapter 5. Following [2.7], let us consider the evolution of the SG equation soliton in a medium with slowly changing parameters.

Let a single SG soliton be given at the moment of time $t = 0$. Let us consider the evolution of this solution for $t \to 0$ under the action of perturbation. The initial equation has the form

$$u_{tt} - u_{xx} + \sin u = \varepsilon f(x, t) R[u], \tag{2.12}$$

We choose $f(x, t)$ in the form $f = \cos \omega t$. Adiabatic equations take the forms

$$v = v_0 - \frac{\varepsilon \omega_0^3}{\omega} - J_0 \sin \omega t \tag{2.13}$$

$$\zeta = \zeta_0 + v_0 t - \frac{\sigma \omega_0^3}{2} - J_0(1 - \cos \omega t) - \frac{\varepsilon v_0 \omega_0^2}{\omega} J_1 \sin \omega t, \tag{2.14}$$

where

$$\omega_0 = \sqrt{(1 - v^2)}, \qquad J_n(v) = \frac{1}{4} \int_{-\infty}^{\infty} R[u_s] z^n \operatorname{sech} z \, dz, \quad n = 0, 1.$$

The first-order correction $u^{(1)}(x, t)$ can be written in the following way (for $t < t_p$):

$$u^{(1)}(x, t) = \tfrac{1}{2}[u_\omega(x, t) + u_{-\omega}(x, t)], \tag{2.15}$$

where $u(x, t)$ is

$$u_\omega(x, t) = -\mathrm{e}^{\mathrm{i}\omega t} \int_{-\infty}^{\infty} \mathrm{d}p \frac{A(p)(p^2 - 1 + 2\mathrm{i}p \tanh z)}{p(1 + p^2)} \left[\frac{1 - \exp(\mathrm{i}b(p)t)}{b(p)} \right] \mathrm{e}^{\mathrm{i}az},$$

$$A(p) = \int_{-\infty}^{\infty} \mathrm{d}z R[u_s(z)](p^2 - 1 - 2\mathrm{i}p \tanh z) \mathrm{e}^{-\mathrm{i}az},$$

$$a = \left(vp - \frac{1}{4vp} \right) \omega_0, \qquad b(p) = \omega - \frac{\omega_0}{2}(p + 1/p). \tag{2.16}$$

As follows from (2.16) for

$$\omega^2 < \omega_0^2 = 1 - v^2,$$

$u^{(1)}(x, t)$ contains localized functions. Accordingly, the perturbation variables with frequencies $\omega^2 < \omega_0^2$ do not excite linear waves in the system, they only change the soliton shape in correspondence with perturbation changes. The dynamics of such a localized perturbation we defined by the adiabatic change of the parameters v and ζ, according to equation (2.13) and (2.14). At $\omega^2 = \omega_0^2$ the correction $u^{(1)}(x, t)$ increases with time as $\sqrt{t}$, i.e. the resonance condition is realized. At $\omega^2 > \omega_0^2$ the linear waves in the system are being excited. This process is comprehensively considered in Chapter 3.

The solution of the equation (2.14) for $\omega^2 > \omega_0^2$, with an accuracy of the order of ε at large times ($t \gg t_0$), takes the form

$$u(x, t) = u_s(z) + u^{(1)}_{\text{loc.}} + u^{(1)}_{\text{rad.}}, \tag{2.17}$$

where u_s is the adiabatic solution with parameters v and ζ described by equations (2.13) and (2.14); $u^{(1)}_{\text{loc}}$ is the localized correction to the soliton shape:

$$u^{(1)}_{\text{loc.}}(z, t) = \frac{\varepsilon\omega_0}{2\omega \cosh z}\left[v(z_0 J_0 - J_1)\sin \omega t + \frac{\omega_0 J_0}{\omega}\cos \omega t \right], \tag{2.18}$$

and $u^{(1)}_{\text{rad}}$ is the propagating linear wave.

One can be convinced that the adiabatic summands are completely compensated by the localized correction $u^{(1)}_{\text{loc}}$ with an accuracy of order ε. So the solution of the initial problem for $t \gg t_0$ with an accuracy up to order ε, is

$$u(x, t) = u_s\left[\frac{x - v_0 t - \zeta + \Delta\zeta}{\sqrt{(1 - v^2)}}\right] + u^{(1)}_{\text{rad}}. \tag{2.19}$$

In the case of even perturbations ($J_0 = 0$), only the constant phase shift appears,

$$\Delta\zeta = \varepsilon\omega_0^3 J_0/4\omega^2.$$

Let us report the results for $R = \sin u$. In the case of a resting soliton ($v = 0$) we obtain

$$u(x, t) = u_s(x, t) + \frac{\pi\varepsilon\sqrt{(k^2 + \tanh^2 x)}}{2k \cosh(\pi k/2)}\sin[kx + \tan^{-1}(\tanh x/k)],\ k = \sqrt{(1 - \omega^2)}. \tag{2.20}$$

This solution describes the excitation of the spin wave in the background of the domain wall (kink).

Let us also study soliton scattering on the localized spatial inhomogeneity

$$f(x) = (\pi l)^{-1/2} \exp[-x^2/l^2], \tag{2.21}$$

and choose $R(u)$ in the form of an odd function. Let us define the change of soliton parameters in the adiabatic approximation. The change of the soliton velocity is found as

$$\Delta v = \int_{-\infty}^{\infty} \frac{dv}{dt} dt = -\frac{\varepsilon}{4v}(1 - v^2)^{3/2} J_0 \int_{-\infty}^{\infty} f(x)\, dx,$$

and the change of the soliton phase as

$$\Delta\zeta = \frac{\varepsilon}{4}\left(\int_{-\infty}^{\infty} f(x)\, dx\right)(1 - v^2) J_1.$$

In the case when R is an odd function, one has

$$\Delta v = 0, \qquad \Delta\zeta = -\varepsilon/4(1 - v^2) J_1. \tag{2.22}$$

It should be noted that the SG soliton evolution in the inhomogeneous medium can be studied without using perturbation theory. As shown in Ref. [2.33], useful arguments are based on the conservation law. For example, consider following this work the perturbation $R[u] = \varepsilon(x) \sin u$, where $\varepsilon(x)$ is a smooth function of the coordinate and $\varepsilon = 0$ at $x = -\infty$ and $\varepsilon = a$ at $x = +\infty$. Defining the energy H for the SG equation as

$$H = \int_{-\infty}^{\infty} \{\tfrac{1}{2}(\varphi_x^2 + \varphi_t^2) + [1 + \varepsilon(x)](1 - \cos\varphi)\}\, dx, \tag{2.23}$$

we have $dH/dt = 0$ for this perturbation. We easily obtain from (2.23) the equality

$$\begin{aligned} \gamma^2(v_f) &= \gamma^2(v_i)/(1 + a) \qquad \text{for} \qquad v_i > [a/(1 + a)]^{1/2}, \\ \gamma(v) &= (1 - v^2)^{-1/2} \end{aligned} \tag{2.24}$$

The comparison with a numerical solution showed that (2.24) is very accurate. In particular, the soliton is expected to reflect from smooth inhomogeneities for $a = 2$.

2.2 THE PROPAGATION OF THE NLS SOLITON IN INHOMOGENEOUS MEDIA

In this section we consider the NLS soliton dynamics in a weakly inhomogeneous medium without dissipation. It should be noted that an exactly solvable NLS model exists for the inhomogeneous medium (Chen and Liu) [2.8]. This model describe nonlinear Langmuir wave propagation in plasma with linearly growing density and the propagation of an intense plane electromagnetic beam in the nonlinear Kerr medium with linearly growing transverse component of the refracton index. In dimensionless variables this perturbed NLS takes the form

$$\mathrm{i}q_t + q_{xx} + 2|q|^2 q = \varepsilon\alpha(t)xq. \tag{2.25}$$

We will search for a solitonic solution of this equation in the form

$$q_\mathrm{s} = 2\eta \operatorname{sech} 2\eta(x - \xi(t)) \exp\{\mathrm{i}B(t)x + \mathrm{i}C(t)\}, \tag{2.26}$$

where

$$\xi = -2\varepsilon\int_0^t\int_0^{t'} \alpha(t'')\,\mathrm{d}t'\,\mathrm{d}t'', \quad \xi B(t) = \varepsilon\int_0^t \alpha(t')\,\mathrm{d}t', \quad C(t) = 4\eta^2 t - \int_0^t B^2(t')\,\mathrm{d}t'.$$

It is easy to see that the influence of nhomogeneities leads only to the modulation of the soliton velocity. At constant gradient ($\alpha < 0$) the soliton is decelerated, and returns back. The change of variables turns the perturbed equation into an ordinary integrable NLS. This result has application to the description of spatial soliton propagation in a periodically modulated medium. The equation for the electric field envelope is

$$2\mathrm{i}kE_z + E_{xx} + k^2(\alpha/2n_0)|E|^2 E = k^2 x\varepsilon(z)E,$$
$$\varepsilon(z) = \varepsilon_0 \cos(z/d_0).$$

Changing the variables in this equation according to

$$q = k\sqrt{(\alpha/n_0)}E, \quad t = z/2k, \quad t_0 = d_0/2k, \quad \varepsilon_0 = a/k^2,$$

we have equation (2.25) with the perturbation $R = a\cos(t/t_0)xq$. Using the result of (2.26) we obtain an equation for the intensity when an electromagnetic beam (spatial soliton) is periodically modulated:

$$|q|^2 = 4\eta^2 \operatorname{sech}^2 2\eta[x - x_0 - vt + 4at_0^2 \sin(t/2t_0)].$$

The modulation period is equal to $\lambda = 2\pi d_0 \sqrt{(1 + v^2)}$. Analogously one can consider dynamics in the nonstationary medium described by the NLS equation with time-dependent coefficients [2.9]. The equation is

$$\mathrm{i}q_t + \beta(\varepsilon t)q_{xx} + 2\alpha(\varepsilon t)|q|^2 q = 0. \tag{2.27}$$

After the variables have been transformed according to

$$t' = 2\int_0^t \beta(\varepsilon t)\,\mathrm{d}t, \quad u(x, t') = (\sigma/2)^{1/2} q(x, t), \quad \sigma = \alpha/\beta, \tag{2.28}$$

equation (2.27) takes the form

$$\mathrm{i}u_t + \tfrac{1}{2}u_{xx} + |u|^2 u = \mathrm{i}\varepsilon\gamma(\varepsilon t)u, \tag{2.29}$$

where $\varepsilon\gamma$ is defined by

$$\varepsilon\gamma(\varepsilon t') = \sigma_{t'}/2\sigma = \tfrac{1}{2}(\ln\sigma)'. \tag{2.30}$$

As can be seen from this equation, the influence of nonstationarity of this form leads to an effective damping or amplification of the soliton. To the first order in ε, the perturbation analysis gives the result for the soliton amplitude the following result:

$$\eta = \eta_0(\sigma/\sigma_0). \tag{2.31}$$

In the case of inhomogeneities of the general form some analytical results may be obtained for the *slowly changing* inhomogeneities [2.10]. The modified NLS has the form

$$\mathrm{i}q_t + q_{xx} + 2|q|^2 q = U(x)q. \tag{2.32}$$

The Lagrangian is

$$L = \frac{\mathrm{i}}{2}\int_{-\infty}^{\infty} (q^* q_t - q_t^* q)\,\mathrm{d}x - \frac{1}{2}\int_{-\infty}^{\infty} (q_x^* q_x - (q^* q)^2 + 2U(x)q^* q)\,\mathrm{d}x. \tag{2.33}$$

As follows from (2.27) equations for the energy E and momentum P take the forms

$$E = \frac{1}{2}\int_{-\infty}^{\infty} [q_x^* q_x - (q^* q)^2 + 2U(x)q^* q]\,\mathrm{d}x,$$

$$P = -\frac{\mathrm{i}}{2}\int_{-\infty}^{\infty} (q^* q_x - q_x^* q)\,\mathrm{d}x. \tag{2.34}$$

For this perturbation the energy E and quantum number $N = \int |q^2 \mathrm{d}x$ are integral invariants. The momentum P is not conserved, but the following equation for P is valid:

$$\frac{\mathrm{d}P}{\mathrm{d}t} = \int_{-\infty}^{\infty} U(x) \frac{\partial}{\partial x} |q|^2 \, \mathrm{d}x = - \int_{-\infty}^{\infty} \frac{\partial U}{\partial x} |q|^2 \, \mathrm{d}x. \tag{2.35}$$

Here it is assumed that the soliton field vanishes at x-infinity: $q(x) \to 0$ as $|x| \to \infty$. Let us consider slowly varying inhomogeneities (in comparison with solitonic scale), when $l_s \gg L$, where L is the inhomogeneity scale, l_s the soliton width.

Then from (2.35), to the leading order in l_s/L it follows that

$$\frac{\mathrm{d}P}{\mathrm{d}t} = - \frac{\partial U(x_0)}{\partial x_0} N, \tag{2.36}$$

where x_0 is the soliton center position.

This equation can be written in the form of the motion equation for a classical particle. Indeed, we represent the one-soliton solution in the form

$$q(x, t) = Q(x - x_0, t) \exp[\mathrm{i}p(x - x_0) + \mathrm{i}S], \tag{2.37}$$

equation (2.32) then shows that (2.37) will be a solution if $\mathrm{d}q/\mathrm{d}t = p$. Substituting (2.37) into (2.32), we obtain

$$P = pN. \tag{2.38}$$

Equations (2.36) and (2.38) indicate that the soliton center moves as a unit mass classical particle in the weakly inhomogeneous potential field $U(x_0)$, with

$$\frac{\mathrm{d}^2 x_0}{\mathrm{d}t^2} = - \frac{\partial U}{\partial x_0}. \tag{2.39}$$

Let the function Q have the form

$$Q_s(x, t) = \sqrt{(2\omega)} \operatorname{sech}(\sqrt{(2\omega)} x) \exp(\mathrm{i}\omega t), \tag{2.40}$$

$$\omega = (N/2)^2,$$

then equation (2.39) coincides with the equation obtained from the adiabatic approximation of IST-based perturbation theory for solitons (see Section 1.).

2.2.1 Simple Examples

(a) *δ-like inhomogeneity.* $V(x) = -\varepsilon\delta(x - x_0)$.
From equation (2.36) we have the effective potential in the form

$$U(x_0) = 8\varepsilon\eta^3 \operatorname{sech}^2 2\eta x_0 \tag{2.41}$$

The possibility of transmission and reflection of solitons, and of oscillating regimes, depends on the initial energy of the solitons. Integration of (2.39) yields the expression for soliton trajectory, for $\varepsilon < 0$, as

$$y(t) = \sinh^{-1}\left\{\sqrt{\left(\frac{2|v_0|}{|E|} - 1\right)}\sin[2|E|(t - t_0)]\right\}, \tag{2.42}$$

$$v_0 = -4\varepsilon\eta^3, \qquad y = 2\eta x_0.$$

Phase plane trajectories for (2.39) are shown in Figure 2.1. The case $E < 0$ corresponds to soliton oscillations on impurities, $E \geqslant 0$ to the unbounded motion.

To investigate soliton oscillations in the δ-like potential we use the action angle variables (I, θ). For the action we have

$$I = (1/2\pi)\oint p\,\mathrm{d}q = (\sqrt{(2|v_0|)} - \sqrt{2|E|}). \tag{2.43}$$

The value of the action upon the separatrix t_0 equals

$$I = I_s - \sqrt{(2|E|)}. \tag{2.44}$$

Applying (2.43), we find the cyclic frequency of soliton oscillations as

$$\omega = \partial E/\partial I = \sqrt{(2|E|)} = (I_s - I). \tag{2.45}$$

As $I \to I_s$, $\omega(I) \to 0$.
(b) *The soliton motion in the periodic potential*: $R[q] = \mathrm{i}\varepsilon\cos(ax)q(x,t)$ [2.11].

Using the adiabatic equations of perturbation theory for NLS solitons (see Appendix 1), we obtain the following set of equations for the soliton parameters:

$$\frac{\mathrm{d}\xi}{\mathrm{d}t} = -\varepsilon\sin(a\zeta)A(a\eta), \quad A^{(a,\eta)} = \frac{\pi a^2}{8\eta\sinh\left(\dfrac{\pi a}{4\eta}\right)}$$

$$\frac{\mathrm{d}\eta}{\mathrm{d}t} = 0, \qquad \frac{\mathrm{d}\zeta}{\mathrm{d}t} = 4\xi, \tag{2.46}$$

$$\frac{\mathrm{d}\delta}{\mathrm{d}t} = 2,4(\xi^2 + \eta^2) + \frac{2\varepsilon\cos a\zeta A(a,\eta)}{a^2}\left[1 - \frac{2}{a}\cosh\left(\frac{\pi a}{4\eta}\right)A(a,\eta)\right],$$

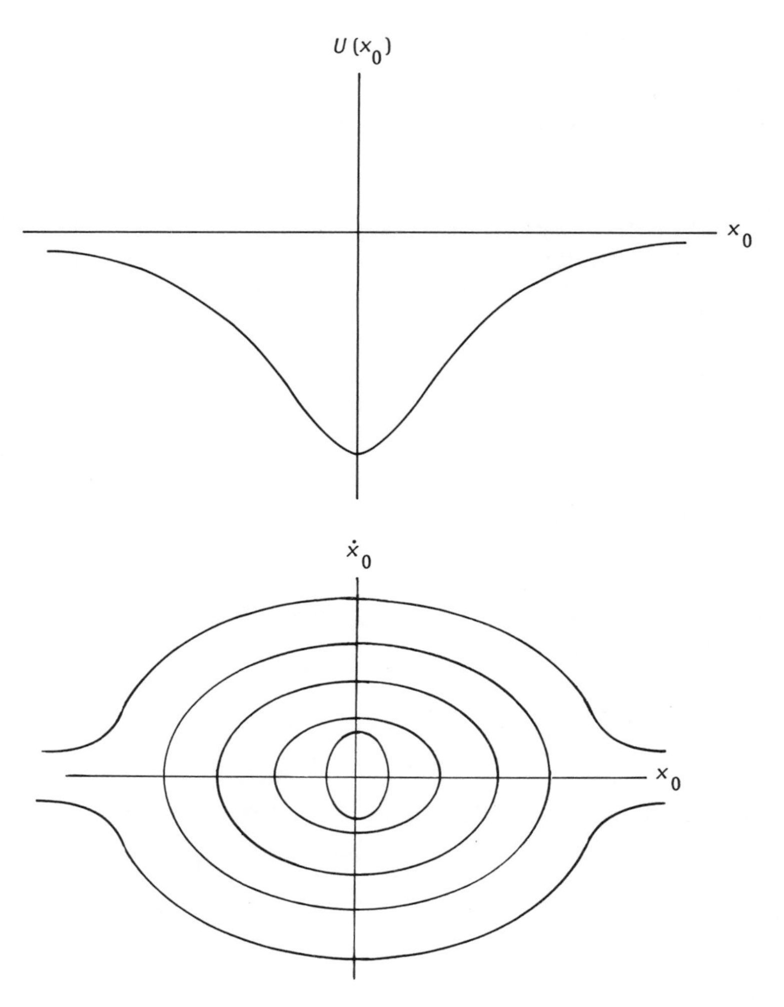

Figure 2.1. Phase plane trajectories for equation (2.2.15)

As a result we obtain a single second-order differential equation for the soliton center coordinate $\zeta(t)$,

$$\frac{d^2\zeta}{dt^2} + 4\varepsilon\eta \sin(a\zeta)A(\zeta, \eta) = 0. \tag{2.47}$$

Here transmitting and trapping regimes are possible. We rewrite this equation in the form of two first-order equations,

$$\dot{p} = -F \sin a\zeta, \qquad \dot{\zeta} = p, \quad F = 4\varepsilon A(a, \eta).$$

The total Hamiltonian is

$$H = \tfrac{1}{2}p^2 - F\cos a\zeta. \tag{2.48}$$

The separatrix is defined by the condition

$$E = F = E_{sx}, \qquad \text{i.e.} \quad E_{sx} = aA(\eta, a). \tag{2.49}$$

When $E > aA$ the soliton motion is unbounded, if $E < aA$, it is finite. The energy of the NLS soliton propagating in the periodic medium is

$$\Delta E_s = \varepsilon \int_{-\infty}^{\infty} \cos(ax)|u_s|^2 \mathrm{d}x = \varepsilon \frac{\pi a \cos(a\zeta)}{\sinh(\pi a/2\eta)}. \tag{2.50}$$

This is the correction to the soliton energy which is given by the periodic medium inhomogeneities; the total energy is

$$E = E_0 + \Delta E_s, \qquad E_0 = 16(\xi^2\eta - \tfrac{1}{3}\eta^3), \quad \zeta = -4\int_0^t \xi(t)\,\mathrm{d}t. \tag{2.51}$$

The frequency of small oscillations near the bottom of the effective potential well is

$$\omega_0 = \sqrt{(aA)}.$$

For nonlinear oscillations $p = \mathrm{cn}(x)$, the frequency is

$$\omega(k) = \pi\omega_0/2K(k), \qquad k^2 = (1 + E/aA)/2.$$

2.3 SOLITON MOTION IN ACTIVE MEDIA

One of the most important cases of soliton dynamics in inhomogeneous media is the soliton propagation in weakly nonconservative media [2.12], [2.13]. To investigate the dynamics of solitons in amplifying media is of special interest. Such kinds of problem appear in hydrodynamics, radiophysics, nonlinear fiber optics, etc. Here we will consider two typical cases: dynamics of KdV solitons interacting with parametrically acting wave pumping, and the NLS soliton propagation in an active inhomogeneous medium. The first problem occurs, for example, in long electromagnetic lines with periodic pumping [2.12], the second one in the propagation of optical solitons in the long optical fibers with amplification [2.14].

Let us consider the amplification of KdV solitons interacting with a periodic parametrically acting wave. One can also assume that weak high- and low-frequency damping exists in the medium. Then the modified KdV equation has the form [2.12]

$$U_t + UU_x + U_{xxx} = -\delta U + \beta U_{xx} + f(\eta), \tag{2.52}$$

where $\eta = x - wt$. We assume that equation (2.52) has a solution in the form of a periodic wave $\tilde{U} = \tilde{U}(x - wt)$, the stationary form of which is supported by the external force $f(\eta)$. The problem of finding this solution is reduced to solving ordinary differential equations of well-known kind (see for example [2.15]). From this point of view we will assume that this part of the problem is solved, and the function $f(z)$ is given.

Following Gorshkov's work [2.16] we will study the evolution of solitary-like perturbations on this periodic wave background. From equation (2.52) we derive the equation for perturbations:

$$U_t + (U + \tilde{U})U_x + U_{xxx} = -\delta U + \beta U_{xx} - \tilde{U}_x U, \tag{2.53}$$

where it is assumed that $U \sim \tilde{U}$, but that the scales of the fields U, $\tilde{U}$ are essentially different, i.e. $\tilde{U}_{x,t} \sim \varepsilon U_{x,t}$, $\varepsilon \ll 1$. Then all the terms in the left-hand side of (2.53) are small. In this case the perturbation method can be applied to the investigation of the soliton evolution. For this purpose we represent the solution in the form of an expansion (1.70), where the first term is

$$U^{(0)} = 3(v - \tilde{U})\,\mathrm{sech}^2[\tfrac{1}{2}\sqrt{(v - \tilde{U}\zeta)}], \qquad \zeta = x - vt. \tag{2.54}$$

Using the direct perturbation method described in Section 1.8, we obtain the equation for velocity $v(t)$:

$$\frac{\mathrm{d}v}{\mathrm{d}t} + w\tilde{U}'_\eta - \tfrac{1}{3}(v - \tilde{U})\tilde{U}'_\eta - \tilde{U}\tilde{U}_\eta + (\tfrac{4}{3})\delta(v - \tilde{U}) + \tfrac{4}{15}\beta(v - \tilde{U})^2 = 0. \tag{2.55}$$

Here $\tilde{U}(\eta)$ and $\tilde{U}'_\eta$ are calculated at $x = \int^t \mathrm{d}t' v$. Introducing the variables

$$B(t) = v - \tilde{U}(\varphi); \qquad \varphi(t) = \int_0^t \mathrm{d}t(v - w), \tag{2.56}$$

we obtain a set of first-order equations:

$$\frac{\mathrm{d}B}{\mathrm{d}t} = -\tfrac{2}{3}B(\tilde{U}' + 2\delta) - \tfrac{4}{15}\beta B^2$$

$$\frac{\mathrm{d}\varphi}{\mathrm{d}t} = B + \tilde{U}(\varphi) - w, \tag{2.57}$$

where $\mathrm{d}B/\mathrm{d}t = \mathrm{d}v/\mathrm{d}t - (v - w)\tilde{U}'(\varphi)$ is the total derivative of the soliton amplitude with respect to time and $\varphi(t)$ is the soliton phase defining the soliton position with respect to the periodic wave. This system can be obtained also by the IST-based perturbation method in the limit of large-scale inhomogeneities, when the soliton length satisfies $l_s \ll L_u$ [2.17].

For $\delta = \beta = 0$ and with an arbitrary ratio of fields $\tilde{U}$ and U_0, the set (2.57) represents a Hamiltonian system. The corresponding Hamilton equations are

$$\dot{I} = -\frac{\partial H}{\partial \varphi}; \qquad \dot{\varphi} = \frac{\partial H}{\partial I}, \tag{2.58}$$

and the Hamiltonian is

$$H = I\tilde{U}(\varphi) + \tfrac{3}{5}I^{5/3} - wI; \qquad I = B^{3/2}. \tag{2.59}$$

In this case (2.57) can be solved. Here, as in the case of a classical particle interacting with the field of a periodic wave, 'passing' and 'trapping' solitons exist.

For nonvanishing dissipation ($\delta \neq 0, \beta \neq 0$) the analysis can be performed with the given form of $\tilde{U}(\eta)$. Consider the case when $\tilde{U}(\varphi) = m\cos(k\varphi)$, where m and k are the amplitude and wave number of the periodic wave. From (2.58) we derive

$$\begin{aligned} \frac{\mathrm{d}B}{\mathrm{d}t'} &= -\tfrac{2}{3}B(-m\sin\varphi' + 2\delta') - \tfrac{4}{15}\beta' B^2, \\ \frac{\mathrm{d}\varphi'}{\mathrm{d}t'} &= B + m\cos\varphi' - w, \end{aligned} \tag{2.60}$$

where new variables are defined by

$$\delta' = k\delta, \quad \beta' = k\beta, \quad \varphi' = k\varphi, \quad t' = t/k.$$

This system is close by the form of equations describing charged-particle dynamics in a ring accelerator [2.18]. At $m < 2\delta'$, solitons are damped. In the opposite case small-amplitude solitons in the interval

$$\pi/2 - \sin^{-1}(2\delta'/m) < \varphi < \pi/2 + \sin^{-1}(2\delta'/m)$$

are amplified. Investigation shows that the system lacks closed integral curves, i.e. nontrivial periodic solutions are absent. The equilibrium (steady-state) points of this system may be found. They are

$$B_{1,2}=\frac{w-\frac{4}{3}\beta'\delta'\pm[(1+\frac{4}{25}\beta'^2)m^2-4(\delta'+\frac{1}{5}\beta'w)^2]^{1/2}}{1+\frac{4}{25}\beta'^2};$$
$$\sin(\varphi'_{1,2}+\varphi^*)=\cos\varphi^*,\qquad \tan\varphi^*=\tfrac{2}{5}\beta'm,\quad B_{3,4}=0. \tag{2.61}$$

Here one can isolate four regions in the plane (m, w), in which the system behaves differently.

(a). The region where $-\sqrt{(m^2-4\delta'^2)}<w<\sqrt{(m^2-4\delta'^2)}$. In this case the points $B=0$, φ' and $\varphi_4=\cos^{-1}(-w/m)$ corresponding to the synchronous motion of small-amplitude solitons lie in the above-mentioned region. The small-amplitude solitons are amplified to the steady state value $B=B_1$.

(b).
$$\sqrt{(m^2-\delta^2)}<\tfrac{5}{2}(\beta')^{-1}(m\sqrt{(1+\tfrac{4}{25}\beta'^2)}-2\delta')\qquad \text{if } w<\tfrac{4}{5}\beta'\delta'$$
$$<\frac{2\delta'}{\frac{2}{5}\beta'+\sqrt{(1+\frac{4}{25}\beta'^2)}} \tag{2.62}$$
$$\sqrt{(m^2-\delta'^2)}<w<m\qquad \text{if } w>\frac{2\delta}{\frac{2}{5}\beta'+\sqrt{(1+\frac{4}{25}\beta'^2)}}$$

The point $B=0$, $\varphi'=\varphi'_4$ are found beyond the amplification region. Then the small-amplitude solitons are damped, and those having sufficiently large amplitudes are amplified to the value $B=B_1$.

(c). $m<w, \frac{5}{2}(\beta')^{-1}(m\sqrt{(1+\frac{4}{25}\beta'^2)}-2\delta')$,

$$w>\frac{2\delta'}{\frac{2}{5}\beta'+\sqrt{(1+\frac{4}{25}\beta'^2)}}.$$

In this case a synchronous motion of small-amplitude solitons together with the periodic wave is impossible, because there is a nonequilibrium point in the $B=0$ axis. A rigid amplification regime takes place.

(d). At $w>\frac{5}{2}(\beta')^{-1}(m\sqrt{(1+\frac{4}{5}\beta'^2)}-2\delta')$, all the solitons are damped and the system has no steady states.

The process of establishing a steady state corresponding to a soliton with constant B and φ' is associated with desynchronization: the change of the soliton amplitude and phase leads to detuning between the soliton and periodic wave motion. As a result, the soliton propagates along the wave up to the point at which there is balance between the energy transferred to the soliton by the wave and the damping loss. The mechanism is analogous to that of collecting particles in ring accelerators [2.18].

Different mechanisms for soliton amplification in inhomogeneous media include the envelope soliton propagation in the long optical fiber system with periodically located amplifiers. This problem has been considered by many authors [2.14], [2.19]. Below we follow Kodama and Hasegawa [2.14].

The equation describing this problem is

$$q_x + q_{\tau\tau} + 2|q|^2 q = \mathrm{i}\varepsilon(x) q(1 - \alpha|q|^2) - \mathrm{i}\gamma q, \tag{2.63}$$

where

$$\varepsilon(x) = \varepsilon_0 T \sum_{-\infty}^{\infty} \delta(x - nT).$$

Here α is the nonlinear dissipation coefficient. It should be noted that x denotes the propagation distance along the fiber, τ is the dimensionless time in the moving reference frame, i.e. $\tau = t - x/v_{\mathrm{g}}$, where v_{g} is the group velocity. We will now describe the evolution of the single envelope-soliton in amplifying media. Substituting the soliton expression (1.17) into the adiabatic equations, we obtain equations for soliton parameters (Kodama, 1985) as

$$\Delta\eta_i = 2\varepsilon_0(1 - \tfrac{2}{3}\alpha\eta_i^2)\eta_i. \tag{2.64}$$

As a result we have the one-dimensional map for η_i

$$\eta_{i+1} = \exp(-2\Gamma T)[1 + 2\varepsilon_0(1 - \tfrac{2}{3}\alpha\eta_i^2)]\eta_i.$$

Here there is a fixed point η_* for $\varepsilon_0 > [\exp(2\Gamma T) - 1]/2 \sim \Gamma T$, given by

$$\eta_*^2 = \frac{3}{2\alpha}\left[1 - \frac{\exp(2\Gamma T)}{1 + 2\varepsilon_0}\right]. \tag{2.65}$$

Then a soliton with amplitude equal to η_* propagates in the optical fiber. It is clear that the velocity of solitons in the adiabatic approximation is unchangeable. The soliton amplitude and phase change periodically (when $\gamma = 0$). Here different regimes are possible, when the scale differs from L_{d}, or is of the same order. When the soliton scale is small as compared with the L_{d}, one can apply this adiabatic method. Otherwise, it is necessary to use another approach. Hasegawa and Kodama use an averaging method excluding the rapidly oscillating terms [2.19].

The propagation of envelope solitons in an inhomogeneous slowly amplifying medium may be also studied in a similar way [2.32]. With an inhomogeneity-amplifying function that is linear in the wave field $q(x, t)$, the perturbed NLS equation has the form

$$iq_t + q_{xx} + 2q^2q^* = \mp i\varepsilon(x)q. \tag{2.66}$$

The sign $(+)$ corresponds to amplification, the sign $(-)$ to energy dissipation. Here $\varepsilon(x)$ is a slowly varying function in comparison with the solitonic scale $(\sim \eta^{-1})$. In Ref. [2.32] the adiabatic set of equation for soliton parameters was obtained and solved.

The third example is the SG small amplitude breather dynamics in an active medium. The corresponding modified SG equation has the form

$$u_{tt} - u_{xx} + \sin u = \varepsilon(t)\sin u - \Gamma u_t. \tag{2.67}$$

This equation appears in the problem of domain wall dynamics in liquid crystals in varying electric and magnetic fields, [2.20], and soliton dynamics in quasi-one-dimensional magnets under external (constant and oscillating) weak fields, [2.21]. We study the behavior of a single small amplitude breather. As is known (see Appendix 2) the breather solution of the unperturbed SG equation is described by

$$u_B(x, t) = -2\tan^{-1}\left(\frac{v}{\eta}\frac{\cos\varphi}{\cosh z}\right), \tag{2.68}$$

$$\psi = \varphi(\tau) - \frac{\eta}{v}zv, z = \frac{v}{|\lambda|}\frac{\xi - \xi_0}{\sqrt{(1 - v^2)}},$$

$$v = \frac{1 - 4|\lambda|^2}{1 + 4|\lambda|^2}, \qquad \lambda = \eta + iv \quad (\eta > 0, v > 0). \tag{2.69}$$

Unlike the kinks, breathers exhibit oscillations of frequency $d\varphi/dt$. The amplitude of the breather is $\gamma = \tan^{-1}(v/\eta)$. For $\gamma \to \pi/2$ the breather may be regarded as the superposition of a weakly bound soliton and an antisoliton; for $\gamma \to 0$ it has a small amplitude. Following the perturbation theory for breathers, [2.22], one can derive equations for γ and φ. They have a rather cumbersome form, so their analysis is difficult. Note that the velocity of the breather, for both variants of the time-varying field, is constant. Here we consider the case when the small-amplitude breather field switching is constant. We have the following set of equations for γ

and φ:

$$\frac{d\gamma}{dt} = 2\gamma[2\varepsilon(t)\cos\varphi - \Gamma\sin\varphi]\sin\varphi, \tag{2.70}$$

$$\frac{d\varphi}{dt} = \cos\gamma + [2\varepsilon(t)\cos\varphi - \Gamma\sin\varphi]\cos\varphi. \tag{2.71}$$

To search for a stationary regime, the phase of breather oscillations will be divided into slow and fast parts, $\varphi = \Delta + (\Omega/2)t$. Δ is considered to be a slow function of time. In (2.70–71) the explicit time dependence will be excluded by averaging over the external field period. As a result, a set of equations is obtained for the averaged magnitudes

$$\frac{d\bar{\gamma}}{dt} = \bar{\gamma}(-\Gamma + \varepsilon_0\cos 2\bar{\Delta}),$$
$$\frac{d\bar{\Delta}}{dt} = \cos\bar{\gamma} - \frac{\Omega}{2} - \varepsilon_0\sin 2\bar{\Delta}. \tag{2.72}$$

Equation (2.72) is seen to allow stationary solutions, namely:

$$\bar{\Delta} = \tfrac{1}{2}\cos^{-1}(\Gamma/\varepsilon_0),$$
$$\gamma = \cos^{-1}(\Omega/2 + \sqrt{(\varepsilon_0^2 - \Gamma^2)}). \tag{2.73}$$

Thus, it is concluded that for the case where a parametrically acting time-varying field is applied there exists a breather synchronized with the external time-varying field. The possibility for such a state to exist is associated with the fact that the energy dissipation due to damping is balanced by supply from the external field energy.

2.4 THE INTERACTION OF SOLITONS IN INHOMOGENEOUS MEDIA

We shall consider the interaction of two solitons in inhomogeneous media. This process can be treated as a 'three-particle' process in the case of localized inhomogeneities, because inhomogeneity plays the role of a third particle.

(a). Let us consider, following Kivshar and Malomed [2.23], the interaction of kinks in the SG system with a localized inhomogeneity of the form

$$R[u] = \varepsilon\delta(x)\sin u. \tag{2.74}$$

Below we consider the case of collision between a fast kink and a slow one in the presence of δ-type inhomogeneity. The change of spectral parameters due to perturbation is defined by

$$\frac{d\lambda_n}{dt} = \frac{\varepsilon}{4}\sigma_n\lambda_n \int_{\infty}^{\infty} dx \frac{\sin[2u(x,t)]}{\cosh[z_n + (-1)^n D/2v_n]}, \tag{2.75}$$

and λ_n are expressed in terms of the kink velocities v_n through

$$\lambda_n = \mathrm{i}\nu = \frac{\mathrm{i}}{2}\sqrt{[(1+v)/(1-v)]}. \tag{2.76}$$

On integration we obtain the changes of parameters for fast and slow kinks in the forms

$$\Delta\lambda_3 = -2\varepsilon v_1 \nu_1 \nu_3^{-1} g(\delta), \tag{2.77}$$

$$\Delta\lambda_1 = 2\varepsilon\lambda_1 \nu_3^{-1} g(\delta), \tag{2.78}$$

where $g(\delta) = \tanh\delta\,\mathrm{sech}^2\,\delta$.

The parameter δ characterizes the overlap between the slow soliton and the inhomogeneity during the collision (at $t = 0$), $\delta = \nu_1(x_1^{(0)} - v_1 x_3^{(0)})$. The total energy of the kinks is conserved, i.e.

$$E = 8\sum_{j=1,3}(\lambda_j + \tfrac{1}{4}\lambda_j). \tag{2.79}$$

The momentum P absorbed by the inhomogeneity is

$$\Delta P = 8\varepsilon\nu_1^{-1}(1 - v_3^2)^{1/2} g(\delta). \tag{2.80}$$

Some results may be obtained by considering the motion of the slow soliton as a classical particle in an effective potential $U(\zeta)$. In the case $\varepsilon < 0$, the kink may be captured or 'pinned' by the impurity potential. The attraction potential has the form (see Example 2.2.1)

$$U = -2\varepsilon\,\mathrm{sech}^2\,\zeta, \tag{2.81}$$

and from this equation it immediately follows that the law of kink motion is

$$\sinh\zeta(t) = \sqrt{[2\varepsilon - E)/E]}\sin((\sqrt{E})t/2), \tag{2.82}$$

Here E is the 'pinning' energy, $0 < E \leqslant 2\varepsilon$. The collision of a fast kink with a slow one which is pinned by a soliton leads to expulsion of the slow kink from the bound state. For the estimation of the velocity for the escaping fast kink we use a formula for the collision-induced phase shift $\Delta\zeta$ of the slow kink, which is valid for $\varepsilon = 0$:

$$\Delta\zeta \approx -\ln\left(\frac{1+(1-v_3^2)^{1/2}}{1-(1-v_3^2)^{1/2}}\right), \tag{2.83}$$

where v_3 is the velocity of the fast kink. The collision time $\sim v_3^{-1} \ll T_0 = 4\pi/\sqrt{E} \sim \varepsilon^{-1/2}$, where T_0 is the oscillation period of pinned kink. For this reason one can neglect the change in the slow soliton velocity during the collision. Then the change of the total energy of the slow kink is equal to the change of the potential energy:

$$\Delta U(\zeta) = U(\zeta_0 + \Delta\zeta_0) - U(\zeta_0). \tag{2.84}$$

Using (2.81) and (2.83) and the condition for escape, $\Delta U > E$, we see that slow kink 'depinning' occurs at $v_3 \gg \sqrt{\varepsilon}$. In the opposite case of v_3 small, the analysis shows that the reflection of a kink from a pinned kink is possible. The threshold reflection velocity is $v_{\text{thr}}^2 = \frac{1}{4}E$. The results obtained can be applied to the collision between free and pinned fluxons on microshorts or microresistance in a long Josephson junction [2.23] as well as to designing logic elements of computers based on LJJ elements.

(b) As the second example, the interaction of solitons in the periodic wave field will serve. We will study the KdV case. Most important is the case when the KdV soliton velocities are close to each other. Also it is very important when the soliton–soliton interaction is of the same order as the soliton interaction with periodic wave modulation of the medium. In this case the equation describing the interaction of KdV solitons in the presence of the wave may be obtained from equation (2.57) in the following form [2.16]:

$$\begin{aligned}
\frac{\mathrm{d}B_i}{\mathrm{d}t} &= -\tfrac{2}{3}B_i(\tilde{U}(\varphi_i) - 2\delta) - \tfrac{4}{15}\beta B_i^2 - 32B_i^{5/2} \\
&\quad \cdot[\exp(-\sqrt{B_i}(\varphi_{i-1} - \varphi_i) - \exp[-\sqrt{B_i}(\varphi_{i+1} - \varphi_i)], \\
\frac{\mathrm{d}\varphi_i}{\mathrm{d}t} &= B_i + \tilde{U}(\varphi_i) - w.
\end{aligned} \tag{2.85}$$

Here we take into account only the nearest interactions of solitons. This system is rather too complicated for complete analysis. However, it is possible to find steady states of this system for a small number of solitons. Using the closeness of soliton amplitudes and phases, we can write these in the forms

$$B_i = B_{1,2} + b_i, \quad \varphi_i = \varphi_{1,2} + \Delta\varphi_i, \quad b_i, \quad \Delta\varphi_i \ll B_{1,2}, \quad \varphi_{1,2}.$$

Here $B_{1,2}$ have already been previously found (see (2.61)). Expanding the right-hand side of (2.85) in series in b_i and $\Delta\varphi_i$, we obtain a set of first-order equations:

$$a\Delta\varphi_i = \exp[-\sqrt{B_{1,2}}(\Delta\varphi_{i+1} - \Delta\varphi_i)/k] - \exp[-\sqrt{B_{1,2}}(\Delta\varphi_i - \Delta\varphi_{i-1})/k],$$
$$b_i = -(m \sin\varphi_{1,2})\Delta\varphi_i, \tag{2.86}$$

where

$$a = \pm(B_{1,2})[(1 + \tfrac{4}{25}\beta'^2)m^2 - (2\delta' + \tfrac{2}{5}\beta' w)]^{1/2} \tag{2.87}$$

and the signs $(\pm)$ refer to unstable and stable states, respectively. The states of interest exist only for $a > 0$, i.e. near the existing steady state. Indeed, since

$$\sum_{i=1}^{N} \Delta\varphi_i = 0, \tag{2.88}$$

it follows that

$$\Delta\varphi_1 < 0, \qquad \Delta\varphi_N > 0.$$

As follows from (2.88), for any N this system admits a symmetric solution

$$\Delta\varphi_i = -\Delta\varphi_{N-i-1}.$$

For $N = 2$ there exists a unique solution, where $\Delta\varphi_{1,2}$ are defined by the equation

$$\Delta\varphi_1 = -\Delta\varphi_2 = 2k \ln(-a\Delta\varphi_2)/\sqrt{B}.$$

For $N = 3$, excluding $\Delta\varphi_3$ and $\Delta\varphi_2$ from equation (2.86), we obtain

$$a(\Delta\varphi_1 - k \ln(a\Delta\varphi_1)/\sqrt{B_1}) = (a\Delta\varphi_1)^{-2}[\exp(3/\sqrt{B_1}\Delta\varphi_1/k) - (a\Delta\varphi_1)^3]. \tag{2.89}$$

The root of this equation deriving from the condition

$$\Delta\varphi_1 = k \ln(a\Delta\varphi_1)/\sqrt{B_1},$$

corresponds to the symmetric solution of (2.86).

The possibility of existence of bound steady states is obvious from the physical point of view. Solitons in the group with $B_i > B_1$ (or $B_i < B_1$)

reside in regions where the energy obtained from the periodic wave is lower (or higher) than the energy loss from damping. Due to the interaction of solitons, the energy is redistributed so that the energy balance for each soliton in the group is possible.

The KdV soliton dynamics in media with slowly varying parameters has been investigated [2.36–2.39].

The *generation* of a KdV soliton in the periodic field is also feasible. This question has been comprehensively studied in Gorshkov's thesis [2.16]. Recently the NLS soliton interaction has been studied in [2.40, 2.41].

2.5 THE INTERACTION OF SOLITONS IN MULTICHANNEL SYSTEMS

One of the examples of an inhomogeneous system is a composite system in which low-dimensional nonlinear subsystems may be separated. For example these may be polymers composed of molecular chains, each of which can be considered as a one-dimensional crystal; or they may be magnets or multifiber optical systems, multilayered optical waveguides, etc. In this subsection we shall consider the interaction of solitons in two coupled NLS systems. This system describes the nonlinear dynamics of short electromagnetic pulses in tunnel-coupled waveguides.

In standard dimensionless variables, the equation for the envelopes of the electric fields in two identical fibers with refractive index $n = n_0 + n_2 |E|^2$ (for fibers of fused quartz $n_2 \cong 10^{-22}\ (m/V)^2$) has the form of a pair of coupled NLS equations [2.24],

$$\mathrm{i}q_{nx} + \tfrac{1}{2} q_{n\tau\tau} + |q_n|^2 q_n = \varepsilon q_m, \qquad n, m = 1, 2 \quad (n \neq m), \tag{2.90}$$

where x is the normalized distance of propagation, τ is the normalized travelling time, and ε is the coupling coefficient between the fibers, which can be calculated, by use of the Gaussian approximation for the field of mode, as [2.26]

$$\varepsilon = \frac{\Delta n}{n_2 |E_0|^2} \exp\left(-\frac{l^2}{4\rho_0^2} \right). \tag{2.91}$$

In this expression Δn is the difference between indices of refraction of the fibers and the medium between them, l is the distance between the fibers, ρ_0 is the modal spot size, and E_0 is the initial amplitude of the electric field in the fibers. For this system, the influence of a weak wave in the fiber on a soliton propagating in a neighboring fiber has been analytically investigated in the soliton regime of propagation [2.26]. The propagation of a single soliton pulse in two coupled fibers has been numerically studied in [2.27].

We now investigate the set of equation (2.90) for the case when the initial state in each fiber is a soliton:

$$q_n = 2\eta_n \operatorname{sech}[2\eta_n(\tau - \xi_n)] \exp[2i\mu_n(\tau - \xi_n) + i\delta_n], \quad n = 1, 2, \quad (2.92)$$

where η_n, μ_n, δ_n and ξ_n are parameters characterizing the amplitude, velocity, phase and position of the center of the soliton, respectively. If ε is small, i.e. under weak-coupling conditions, we can use the IST-based perturbation theory for solitons to find the solution of the system [2.25]. For (2.91) the perturbation operator takes the form

$$R_{mn} = \varepsilon q_m, \qquad n, m = 1, 2 \quad (n \neq m).$$

Let us introduce new variables:

$$\eta = (\eta_1 + \eta_2)/2, \quad r = 2\eta(\xi_2 - \xi_1), \quad \psi = \delta_2 - \delta_1,$$
$$\Delta\mu = \mu_2 - \mu_1, \qquad \Delta\xi < \xi_2 - \xi_1.$$

In the adiabatic approximation, we obtain the following equations for the soliton parameters [2.24]:

$$\frac{d\mu_n}{dx} = (-1)^n 2\varepsilon\eta \frac{r\cosh r - \sinh r}{\sinh^2 r} \cos\Psi, \quad (2.93)$$

$$\frac{d\eta_n}{dx} = (-1)^{n-1} 2\varepsilon\eta \frac{r}{\sinh r} \sin\Psi, \quad (2.94)$$

$$\frac{d\xi_n}{dx} = 2\mu_n + \frac{\varepsilon r^2}{2\eta \sinh r} \sin\Psi, \quad (2.95)$$

$$\frac{d\delta_n}{dx} = 2\mu_n \frac{d\xi_n}{dx} + 2(\eta_n^2 - \mu_n^2) + \varepsilon \frac{r}{\sinh r}\left(r\frac{1}{\tanh r} r - 2\right)\cos\Psi. \quad (2.96)$$

In deriving this system we assumed that $\mu_{n0} = 0$, $\eta_{10} = \eta_{20}$, and $\eta_1 \approx \eta_2$. As can be seen from equation (2.94) and (2.96), the assumption about constancy of the amplitude for the interacting solitons holds when the initial phase difference $\Psi_0 = \delta_{20} - \delta_{10} = 0, \pi$.

Now we analyze the set of equations (2.93)–(2.96). From equation (2.93), (2.94) we can prove that the following integrals of motion exist:

$$\mu_1 + \mu_2 = 0, \quad \xi_1 + \xi_2 = \text{constant}, \quad \Psi = \Psi_0. \quad (2.97)$$

Then from equation (2.93), (2.94) we find that

$$\Delta\mu_x = 4\eta\varepsilon \frac{r\cosh r - \sinh r}{\sinh^2 r}\cos\Psi_0, \tag{2.98}$$

$$\Delta\xi_x = 2\Delta\mu. \tag{2.99}$$

Differentiating equation (2.99) with respect to x and taking into account equation (2.98) we obtain a closed equation for r,

$$r_{xx} - 16\eta^2\varepsilon\cos\Psi_0 \frac{r\cosh r - \sinh r}{\sinh^2 r} = 0. \tag{2.100}$$

It follows from equation (2.100) that the potential of the soliton interaction is

$$U = 16\eta^2\varepsilon\cos\Psi_0 \frac{r}{\sinh r}. \tag{2.101}$$

From equation (2.101) and from analysis of equation (2.100) on the phase plane it follows that there exists a bound state of solitons when $\Psi_0 = \pi$. For small $r_0 (r_0 \ll 1)$, equation (2.100) takes the form of the equation of motion for a harmonic oscillator with oscillation period $L = \pi(2\eta^2\varepsilon)^{-1/2}$. For $\Psi_0 = 0$ the effective potential (2.101) turns into a repulsive potential. For $\psi_0 \neq 0, \pi$, the intense energy exchange occurs between solitons and the analysis developed here cannot be applied. This system of equations was also considered numerically [2.27], and the phenomenon of soliton switching was observed.

Interesting also is the case of spatially inhomogeneous coupling between fibers, when $\varepsilon = \varepsilon(x)$. Extending the above-mentioned analysis to this case, we obtain the following equation for the relative soliton distance r, for periodic coupling and $\Psi_0 = \pi$ [2.28]:

$$r_{xx} + 16\eta^2\varepsilon_0\cos(ax)r = 0. \tag{2.102}$$

This is the well-known Mathieu equation. We conclude that with a moving bound-soliton state in a periodically inhomogeneous medium, parametric resonance between the internal oscillations of the bound state and the modulation of the medium is possible [2.29]. The extension of these results to two coupled nonlinear modes (modified Manakov system) was performed in the article [2.30]. An interesting peculiarity is the possibility of exchange between translational energy and internal energy due to pulse-width oscillations. As a result, the interaction between solitons is generally inelastic. When parameter ε is increased, new localized solutions are possible [2.42].

Interactions of solitons in other systems, for example, in shallow stratified liquids, or in two weakly interacting long Josephson junctions are described in the Ref. [2.23].

2.6 PROPAGATION OF KINKS IN INHOMOGENEOUS NONINTEGRABLE SYSTEMS

Here we consider the solitary wave motion in the inhomogeneous nonintegrable systems, and specifically in the so-called φ^4 model. We described this model briefly in Chapter 1 and found solitary wave solutions (kinks) and nonlinear periodic wave solutions. The nonintegrable nature of this model leads to inelastic interactions of kinks, and emission of linear waves in the interaction. The model is important for applications in solid-state physics and in second-order phase transition theory [2.2]. In real systems, however, there always exist various spatial and temporal inhomogeneities. For this reason, we study here the propagation of kinks in media with spatial and temporal inhomogeneities which multiply a linear function of the field. The corresponding equation is

$$u_{tt} - u_{xx} - u + u^3 + \alpha u_t + F(x,t) + V(x,t)u = 0, \tag{2.103}$$

where αu_t represents the dissipation effect and $F(x,t)$ and $V(x,t)$ describe the external field and inhomogeneity effects. Adiabatic dynamics of kinks under the action of additive and multiplicative perturbations were studied in [2.34], [2.35].

Following [2.35] we obtain equations for the kink velocity and the kink energy center. For the unperturbed φ^4 equation a conservation law

$$T_t + X_x = 0 \tag{2.104}$$

is valid. Here T is the conserved density and X is the conserved flux. For example T_1 is the energy and T_2 is the momentum,

$$\begin{aligned} &T_1 = \tfrac{1}{2}(u_t^2 + u_x^2) + \tfrac{1}{4}(u^2 - 1)^2 > 0, \qquad T_2 = u_x u_t. \\ &X_1 = -u_x x_t, \quad X_2 = \tfrac{1}{2}(u_t^2 + u_x^2) - \tfrac{1}{4}(u^2 - 1)^2. \end{aligned} \tag{2.105}$$

Using the results of Chapter 1, we have the following results for kink energy and kink momentum (the kink solution is given by (1.50)):

$$E = \int \mathrm{d}x\, T_1 = \frac{4\gamma}{3\sqrt{2}}, \tag{2.106}$$

$$P = \int_{-\infty}^{\infty} \mathrm{d}x\, T_2 = \frac{4\gamma v}{3\sqrt{2}}, \tag{2.107}$$

$$\gamma = (1 - v^2)^{-1}.$$

Let us define also the kink energy center by the expression

$$x_c = \frac{\int \mathrm{d}x\, x T_1}{\int \mathrm{d}x\, T_1}. \tag{2.108}$$

For the kink solution $X_c = vt + x_0$.

Let us find dynamical equations for the kink parameters in the perturbed case,

$$u_{tt} - u_{xx} - u + u^3 = \varepsilon R[u], \tag{2.109}$$

where $0 < \varepsilon \ll 1$. Differentiating equation (2.106) with respect to t and using equation (2.109), we obtain the differential equation for the velocity of kink.

$$v'(t) = -\varepsilon \tfrac{3}{4}(1 - v^2) \int_{-\infty}^{\infty} \mathrm{d}x \frac{R[u_0]}{\cosh^2 \varphi}, \tag{2.110}$$

where $\varphi = (\gamma/\sqrt{2})[x - z(t)]$. In the same way may find the equation for the kink position $x_0(t)$. Here it is necessary to use the density xT_1. The result is

$$x'(t) = -\varepsilon(3\sqrt{2}/4)v(1 - v^2)^{1/2} \int_{-\infty}^{\infty} \mathrm{d}x \frac{\varphi R[u_0]}{\cosh^2 \varphi}. \tag{2.111}$$

These last equations describe the kink dynamics under perturbations. Consider two particular cases of perturbations:

(1) Let the perturbation R have the form of additive noise with dissipation, i.e.

$$R[u] = -F(x, t) - \alpha u_t, \tag{2.112}$$

where α is a damping constant and $F(x, t)$ is given by

$$F(x, t) = \begin{cases} F(t) & \text{if } x \in [-L, L], \\ 0 & \text{elsewhere.} \end{cases} \tag{2.113}$$

Substituting (2.112), (2.113) into (2.110), (2.111), we obtain the dynamical equations for the kink parameters in the form

$$v'(t) = \varepsilon[-\tfrac{3}{4}\sqrt{2}(1 - v^2)^{3/2}F(t) - \alpha v(1 - v^2)], \tag{2.114}$$

$$z'(t) = v. \tag{2.115}$$

(2) The second case is that of a multiplicative force, when $R[u]$ is equal to $-V(x, t)u - \alpha u_t$. Here $V(x, t)$ satisfies the properties (2.113). Then the equations for kink velocity and position are

$$v'(t) = -\varepsilon\alpha v(1 - v^2), \tag{2.116}$$

$$z'(t) = v + \tfrac{3}{2}\varepsilon v(1 - v^2)V(t). \tag{2.117}$$

3

RADIATIVE EFFECTS IN INHOMOGENEOUS MEDIA

3.1 PROCESSES INDUCED BY SOLITONS CROSSING THE INTERFACE BETWEEN TWO MEDIA

When a soliton crosses the interface between two media, two types of dynamics are possible. The first one is characterized by the change of soliton parameters, and deformation of the soliton profile when pinned on the interface, i.e. it is essentially adiabatic, and can be described in terms of the representation of the motion of a deformed Newton particle in some effective nonlinear potential. The second type of dynamics represents the fact that, while crossing the interface, solitons may radiate plane waves as well as generate new solitons. This possibility is due to the fact that a soliton is a localized wave packet consisting of nonlinearly coupled harmonics, in other words, a packet of coupled quasiparticles. In passing from one medium to another, along with the rearrangement of the wave field, a part of the energy will be released as radiation. This process very much resembles that of transient radiation by charged particles, where part of the intrinsic field is radiated (see review by Feinberg [3.1]). The feasibility of generating new solitons is associated with the fact that the portion of the wave field that has passed the interface may appear to be a sufficient initial condition to generate solitons in a new medium. Examples are provided by the passage of an intense plane beam through the interface, accompanied by a discontinuity in the nonlinear part of the refractive index, and the passage of an ion-sound KdV soliton through a step change in plasma density [3.20].

In this subsection we will consider the adiabatic dynamics and radiative effects describing the intersection of the interface by SG and NLS solitons [3.2], [3.3].

We study the problems of propagation of fluxons through the boundary between two long Josephson junctions in contact, and intense plane electromagnetic beam propagation through the interface.

As the first example we begin with the problem of interaction of an SG soliton (fluxon) with the interface. The corresponding equation describing propagation of the magnetic flux, has the form [3.2]:

$$u_{tt} - u_{xx} + \sin u = \varepsilon R(u), \tag{3.1}$$

$$\varepsilon R(u) = -\gamma u_t - f - \alpha\theta(x)\sin u. \tag{3.2}$$

Here $f = I/I_c$ is the dimensionless bias current, I_c is the critical current in the first junction, $\alpha_c = \Delta I/I_c, \Delta I = I'_c - I_c$ where I'_c is the critical current in the second line, γ is the dissipation coefficient, $\theta(x)$ is the Heaviside function

$$\theta(x) = \begin{cases} 1, & x > 0 \\ 0, & x < 0. \end{cases}$$

Below we study the evolution of a single soliton in the form of (1.26). In the adiabatic approximation, the equations for velocity and the soliton center coordinate are (for perturbation (3.2) (see Section 1.9))

$$\frac{dv}{dt} = -\gamma v(1-v^2) - \frac{\pi\sigma_f}{4}(1-v^2)^{3/2} - \frac{\alpha(1-v^2)^{3/2}}{2\cosh^2[\xi\sqrt{(1-v^2)}]}; \tag{3.3}$$

$$\frac{d\xi}{dt} = v - \frac{\alpha v}{2}(1-v)\left\{1 + \tanh\frac{\xi}{\sqrt{(1-v^2)}} - \frac{\xi/\sqrt{(1-v^2)}}{\cosh^2[\xi/\sqrt{(1-v^2)}]}\right\}. \tag{3.4}$$

This system is too complicated to be exactly analyzed. For small soliton velocities, i.e. $v^2 \ll 1$, this system reduces to the single equation

$$\frac{d^2\xi}{dt^2} + \gamma\frac{d\xi}{dt} = -\frac{dU(\xi)}{d\xi}, \tag{3.5}$$

$$U(\xi) = \tfrac{1}{4}\pi\sigma f\xi + \tfrac{1}{2}\alpha\tanh\xi. \tag{3.6}$$

This equation describes the motion of an effective classical particle with unit mass in an effective anharmonic potential $U(\xi)$. In Figure 3.1, the form of this potential for different values of parameters α and σ_f is shown. It is clear that for $\sigma_f < 0$ and $|f| < 2\alpha/\pi$ the soliton is pinned by the interface. The corresponding frequency of small oscillations at $\gamma = 0$ is

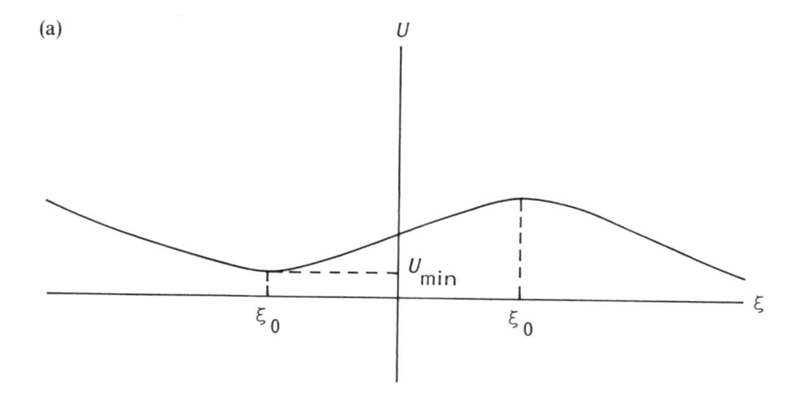

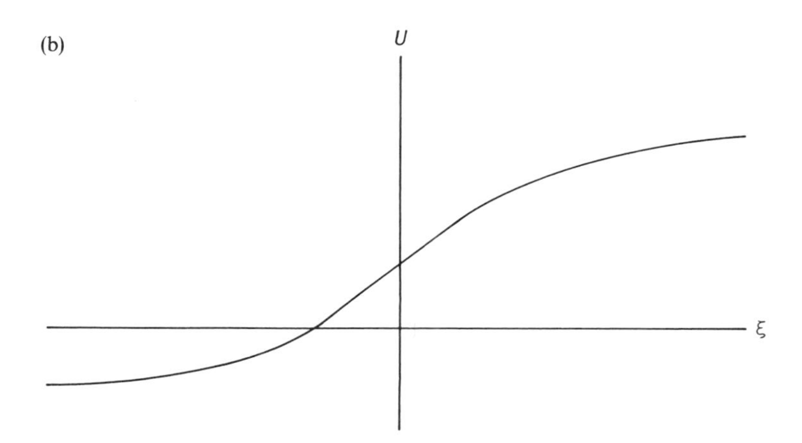

Figure 3.1. Effective potential $U(\xi)$ for equation (3.1.1) at $\alpha > 0$: (a) $\sigma f > 0$, $|f| < 2\sigma/\pi$; (b) $\sigma f > 0$

Ω_0, where

$$\Omega_0^2 = (\pi|f|/16)(1 - \pi|f|/2\alpha)^{1/2}. \tag{3.7}$$

For $\sigma_f > 0$ the soliton is reflected from the interface. We proceed now with the investigation of another phenomenon occurring when a soliton passes through the interface, namely the transient emission of a wave [3.2], [3.4]. Consider the case of a high soliton velocity, when the kinetic energy of the soliton is higher than the change of the potential energy of the soliton in going from one medium to another. In this case the radiation field may be found from perturbation theory. It is useful for this purpose to apply the IST-based perturbation theory (see Appendix 2).

For the first-order correction to the adiabatic approximation we have

$$u^{(1)}(x,t) = -\frac{1}{\pi}\int_{-\infty}^{\infty} \mathrm{d}\lambda \frac{b(\lambda,t)(\lambda^2 - v^2 + 2\mathrm{i}\lambda v \tanh z)}{\lambda(\lambda^2+v^2)} \mathrm{e}^{\mathrm{i}k(\lambda)x}, \tag{3.8}$$

where

$$v = [(1+v)/(1+v)]^{1/2}, \qquad k(\lambda) = \lambda - \frac{1}{4\lambda},$$

and $b(\lambda, t)$ is the first term in the perturbation expansion of the Jost scattering coefficient. The evolution of $b(t)$ is described by the equation

$$\frac{\partial b(\lambda,t)}{\partial t} = -\mathrm{i}\omega(\lambda)b(\lambda,t) - \frac{\mathrm{i}\varepsilon\sqrt{(1-v^2)}}{4(\lambda^2+v^2)}\exp(-\mathrm{i}k(\lambda)\xi)A(\lambda,v),$$

$$A(\lambda,v) = \int_{-\infty}^{\infty} \mathrm{d}z\, R[\varphi_s(z)] \frac{\lambda^2 - v^2 - 2\mathrm{i}\lambda v \tanh z}{\exp[\mathrm{i}k(\lambda)z\sqrt{(1-v^2)}]}. \tag{3.9}$$

The frequency of linear waves is $\omega(\lambda) = \lambda + \frac{1}{4}\lambda$, $\omega^2 = 1 + k^2$. For $t \to -\infty$ the emission is absent, i.e. $b(-\infty) = 0$. Substituting the perturbation (3.2) at $\gamma, f = 0$ we have

$$b(\lambda,t) = \frac{\mathrm{i}\pi\alpha\sigma(1-v^2)}{2v^2}\mathrm{e}^{-\mathrm{i}\omega(\lambda)t}\operatorname{sech}\left[\frac{\pi\sqrt{(1-v^2)}}{2v}\omega(\lambda)\right]. \tag{3.10}$$

Substituting this into (3.9) for the first-order correction describing the emission, we obtain the following expression:

$$u^{(1)}(x,t) = -\frac{\mathrm{i}\alpha\sigma(1-v^2)}{v^2}\int_{-\infty}^{\infty} \frac{k(k - \mathrm{i}v\omega(k)) + \mathrm{i}\sqrt{(1-v^2)}\tanh z}{\sqrt{(1+k^2)}[\omega(k) - kv]}$$
$$\times \operatorname{sech}\left[\frac{\pi\sqrt{(1-v^2)}}{2v}\sqrt{(1+v^2)}\right]\exp\{\mathrm{i}kx - \mathrm{i}\omega(k)t\}. \tag{3.11}$$

The transient emission is the field which propagates far from the interface ($|z| \gg 1$). Then the wave field has the form

$$u^{(1)}(x,t) = \frac{1}{2\pi}\int_{-\infty}^{\infty} \mathrm{d}k\, \Phi(k)\exp[\mathrm{i}(kx - \omega(k)t)],$$

$$\Phi(k) = \frac{[k - v\omega(k) + \mathrm{i}\operatorname{sgn} z\sqrt{(1-v^2)}]}{(\omega(k) - kv)\sqrt{(1+k^2)}}\operatorname{sech}\left[\frac{\pi\sqrt{(1-v^2)}}{2v}\right]\sqrt{(1+k^2)}. \tag{3.12}$$

It is seen that the transient emission from the soliton looks like two wave packets propagating in opposite directions from the interface. Let us find the energy density of the transient emitted waves. As is known, the corresponding density is defined by formula

$$E(k) = \left[E(\lambda) \frac{\mathrm{d}\lambda}{\mathrm{d}k} \right]_{\lambda = \frac{1}{2}[k + \sqrt{(1+k^2)}]} \approx \frac{4}{\pi} b[\lambda(k), t]^2. \tag{3.13}$$

Substituting (3.10) into this equation, we obtain the spectral density of energy for the emitted radiation:

$$E(k) = \sqrt{(2\pi)}\alpha^2 \frac{(1-v^2)}{v^4} \operatorname{sech}^2 \left[\frac{\pi\sqrt{(1-v^2)}}{2v} \sqrt{(1+k^2)} \right] \tag{3.14}$$

The total emitted energy is

$$E_{\text{rad.}} = \int_{-\infty}^{\infty} \mathrm{d}k\, E(k) = 2\sqrt{2}\pi\alpha^2\tau^4 \int_{-\infty}^{\infty} \frac{\omega\, \mathrm{d}\omega}{\sqrt{(\omega^2-1)}} \operatorname{sech}^2 \left[\frac{\pi\tau}{2}\omega \right]. \tag{3.15}$$

The dependence of total energy on velocity is shown in Figure 3.2. At low ($v \to 0$) and high ($v \to 1$) velocities the emission vanishes. In the latter case the soliton form tends to the θ-function form. The part of emission, localized on soliton calculated in [3.26].

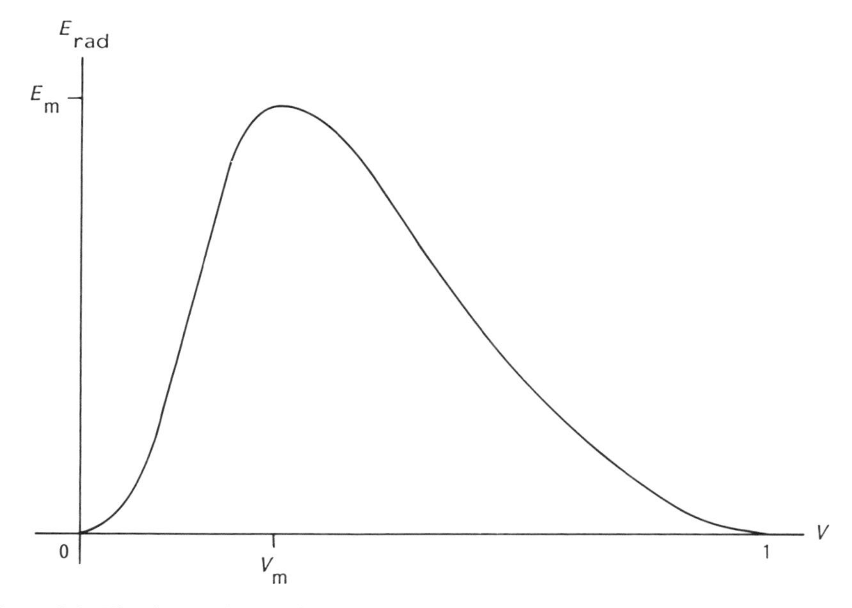

Figure 3.2. The depenedence of total emitted energy on the kink velocity: $E_{\mathrm{m}} \approx 0.13\,\pi\alpha^2$, $V_{\mathrm{m}} \approx 0.67$ [3.2]

The second example is the propagation of an intense electromagnetic beam (the so-called *spatial* solitons) through the interface between two different nonlinear optical media [3.3]. This problem describes also the passage of a small-amplitude SG breather through the interface.

As is known, the equation for the electromagnetic field envelope in a nonlinear dielectric has the form

$$2\mathrm{i}\beta\frac{\partial F}{\partial z}+\frac{\partial^2 F}{\partial x^2}-(\beta^2-n^2)F=0 \tag{3.16}$$

Here the refractive index satisfies

$$n^2=n_{\mathrm{i}}^2+\alpha_{\mathrm{i}}|F|^2, \qquad (\alpha_{\mathrm{i}}>0).$$

This index is assumed to be discontinuous at the interface. For the left-hand side medium ($x<0$) $i=0$, and for the right-hand side $i=1$. Below we will consider the case when

$$\begin{aligned}\Delta &\equiv n_0^2-n_1^2>0\\ \alpha &\equiv \alpha_0/\alpha_1<1.\end{aligned} \tag{3.17}$$

The parameters of the problem are the refractive indices of both media, and those of the incoming electromagnetic beam are its power and the incidence angle. Using the transformation to new variables

$$F(x,z)=\sqrt{\left(\frac{2}{\alpha_0}\right)}\,q(x,t)\mathrm{e}^{\mathrm{i}(\beta^2-n_0^2)z/(2\beta)}, \tag{3.18}$$

where $t=z/2\beta$, we can reduce equation (3.17) to the perturbed NLS equation

$$\mathrm{i}q_t+q_{xx}+2|q|^2q=V(x)q, \tag{3.19}$$

where the perturbation potential has the form

$$V(x)=\begin{cases}0, & x<0\\ \Delta-2(\alpha^{-1}-1)|q|^2, & x>0.\end{cases} \tag{3.20}$$

When there is no interface we have $\alpha=1$, $\Delta=0$, $V=0$. Equation (3.20) can be rewritten in the form $V=\varepsilon(\Delta,\alpha,|q|^2)\theta(x)$.

Let us consider the case of one spatial soliton incident on the interface from the *left*:

$$q(x,t)=2\eta_0\operatorname{sech}2\eta_0(x-\bar{x})\exp\{\mathrm{i}(vx/2+2\sigma)\},$$

where

$$\frac{d\bar{x}}{dt} = v, \qquad \frac{d\sigma}{dt} = -\frac{v^2}{8} + 2\eta_0^2. \tag{3.21}$$

Using the IST-based perturbation theory (Section 2.2), we obtain the following equations for the soliton parameters:

$$\frac{dp}{dt} = 0,$$

$$\frac{d\bar{x}}{dt} = v,$$

$$\frac{dv}{dt} = -2p^{-1}\int_{-\infty}^{\infty} \frac{\partial V(x,t)}{\partial x} q(x,t)q^*(x,t)dx. \tag{3.22}$$

This system is reduced to a single equation for the soliton center coordinate (so-called equivalent particle formalism) [3.3]

$$\frac{d^2x}{dt^2} = -2p^{-1}\int_{-\infty}^{\infty} \frac{\partial V(x,t)}{\partial x} q(x,t)q^*(x,t) = -\frac{\partial U_L(\bar{x})}{\partial \bar{x}}. \tag{3.23}$$

which is the equation of motion of a unit mass classical particle in the anharmonic effective potential

$$U_L(\bar{x}) = \Delta(1-(\alpha S_0)^{-1})\tanh(2\eta_0\bar{x}) + (\Delta(3\alpha S_0)^{-1})\tanh(2\eta_0\bar{x}).$$

The form of this potential is shown in Figure 3.3.

A set of possible trajectories in the phase plane is defined by the expression

$$v^2(\bar{x}) - v_0^2 = 2(U_L(\bar{x}_0) - U_L(\bar{x})),$$

where $(\bar{x}_0, v_0)$ are the initial position and velocity. Here the incidence angle is $\phi_i = \sin^{-1}(v_i/2)$. When the initial velocity v_0 is such that the total energy $E \gg \text{Max}(U_L)$, the spatial soliton crosses the interface. For $\alpha S_0 < 1$ there is a maximum in U_L, corresponding to the existence of the unstable steady surface wave, centered in the left medium. When the spatial soliton crosses the interface, linear waves are emitted. The amount of radiation may be calculated using the IST-based perturbation formula (see Appendix 1)

$$N_{rad} = -\frac{2}{\pi}\int_0^{\infty} \ln(1-|b|)^2\, d\lambda \approx \frac{2}{\pi}\int_0^{\infty} |b|^2\, d\lambda.$$

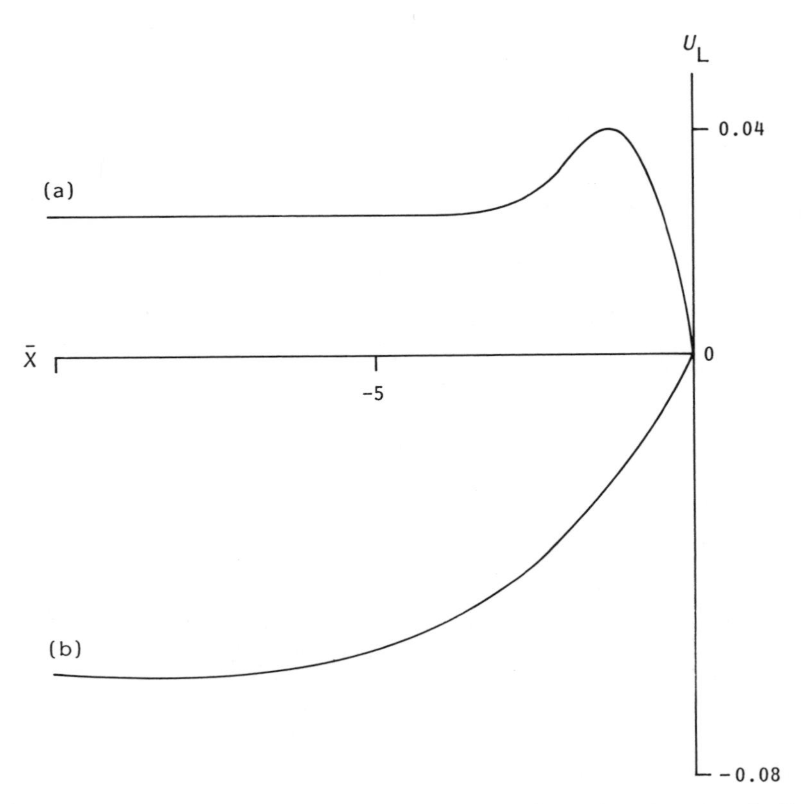

Figure 3.3. Effective potential for the beam center coordinate for $\sigma S_0 < 1$ and $\alpha S_0 > 1$

The calculation shows that the power converted into radiation is

$$N_{\text{rad}} = (1 - \sqrt{a})^2 \frac{8\eta_0}{\alpha_0}.$$

The radiation energy tends to zero as α tends to unity.

The case of modulated interface has been studied in [3.27].

Together with the generation of linear waves in crossings of the interface, it is possible to generate new solitons also. That was predicted for the KdV soliton case by Tappert and Zabusky [3.19], and by Pelinovsky [3.20]. They considered the KdV soliton incident on a gradient region between two uniform media. The basic idea consists in the consideration of the KdV soliton transition from the 1-medium into the second as the initial condition for the 2-medium. Applying the WKB approximation for solving linear spectral problems, these authors derive formulas giving the number and amplitude of the 'fission solitons'. If the initial field has the form

$$u_s = 12(\beta_-/\alpha_-)\kappa^2 \operatorname{sech}^2[\kappa(x - 4\kappa^2\beta_- t)],$$

then the number of solitons generated in the second medium is

$$N = [\tfrac{1}{2}\{\sqrt{(1 + 4(\alpha_+\beta_-)/(\alpha_-/\beta_+))} - 1\}].$$

Here [...] denotes the integer part, and $\alpha_\pm$, $\beta_\pm$ are the value of nonlinear and dispersive coefficients of the KdV equation for $t > t_0$ and $t < t_0$ correspondingly. This approach can also be applied to the NLS equation with step-like coefficients describing, for example, crossing of an interface by optical solitons in an inhomogeneous fiber (Abdullaev *et al.* [3.25]).

3.2 EMISSION OF WAVES BY SOLITON IN PERIODICALLY MODULATED MEDIA

Here we will study the emission of radiation by soliton moving in the periodically modulated medium. The first example is the SG kink motion. In the case of a long Josephson junction with periodically modulated parameters, the corresponding perturbed SG equation for the phase difference is

$$u_{tt} - u_{xx} + \sin u = \varepsilon \sin(kx) \sin u. \tag{3.22}$$

The emission field of a soliton and the emitted energy were found by many authors [3.4]–[3.6]. According to [3.6] we will consider the emission process that takes place under the condition that $v^2 > (1 + \kappa_0^2)^{-1}$, where v is the soliton velocity, κ_0 is the radiation wavenumber.

Let us calculate the energy emitted by a soliton (kink) per unit time. The Jost coefficient for perturbation $R[u]$ is equal to

$$b^{(1)}(\lambda, t) = \tfrac{1}{2}[b_\kappa(\lambda, t) + b_{-\kappa}(\lambda, t)], \tag{3.23}$$

where

$$b_\kappa(\lambda, t) = -\frac{\varepsilon v \lambda A(\lambda, v, \kappa)}{2(\lambda^2 + v^2)} e^{-2ik(\lambda)\zeta} e^{i\kappa\zeta} \frac{1 - e^{-i\varphi(\lambda)(1-v)t/\lambda}}{\varphi(\lambda)}. \tag{3.24}$$

Here $\varphi(\lambda) = \lambda^2 + v^2 + 2v\kappa/\kappa_0\lambda$, κ_0 was defined above, and A is given by

$$A(\lambda, v, \kappa) = \int_{-\infty}^{\infty} dz\, R[u_s(z)](\lambda^2 - v^2 - 2i\lambda v \tanh z) e^{iz[\kappa - 2k(\lambda)]\sqrt{(1-v^2)}}. \tag{3.25}$$

Using formulae (A.62), (3.24) and (3.25) we obtain the expression for the total energy emitted by the system per unit time,

$$W(v)=\frac{\varepsilon(1+v)}{32|\lambda_1-\lambda_2|}\sum_{k=1}^{2}\frac{4\lambda_k^2+1}{\lambda_k(\lambda_k^2+\nu^2)}|A(\lambda_k,\nu,\kappa)|^2. \tag{3.26}$$

For the periodic perturbation $R[u]=\varepsilon\sin(kx)\sin u$, (3.25) takes the form

$$A=\frac{2i\pi\sigma}{\cosh(\pi a/2)}[\lambda\nu(a^2-1)+a(\lambda^2-\nu^2)]. \tag{3.27}$$

Here

$$a=a(\lambda\kappa)=\surd(1-v^2)[\kappa-2k(\lambda)].$$

In the case of small-scale inhomogeneities, when $\kappa\gg1$, and low velocities, $v^2\ll1$, we have, for $\kappa v\gg1$, the estimate

$$W(v)=\frac{\pi\varepsilon\kappa}{4v^2}\operatorname{sech}^2\frac{\pi\kappa}{2}. \tag{3.28}$$

Analogously, for a high velocity, $v^2\to1$, can be obtained

$$W(v)=\frac{\pi^2\varepsilon^2}{4}(1+k^2)(1-v^2)^2. \tag{3.29}$$

The emission by a chain of kinks in periodically inhomogeneous media reveals new features, as compared with the single kink case. In particular, it is possible to obtain the analog of superfluorescence. Let a periodic sequence of microshorts (or microresistances) exist in the LJJ. The corresponding modified SG equation has the form [3.12]

$$u_{tt}-u_{xx}+\sin u=-\gamma u_t-f+\varepsilon\sum_n\delta(x-an)\sin u, \tag{3.30}$$

where f is the d.c. bias current density, a is the spacing between the point-like inhomogeneities. The case $\varepsilon>0$ ($\varepsilon<0$) corresponds to the micro-resistor (or microshort) case. Expressions for the emitted power and the description of superfluorescence phenomena for periodic soliton sequences were given in Ref. [3.5].

Analysis of the NLS soliton emission in a spatially periodic medium can be also performed with the aid the IST perturbation theory for solitons. The model has the following form [3.5]:

$$iq_t+q_{xx}+2|q|^2q=A\cos(kt-kx)q, \tag{3.31}$$

Using the perturbed IST equation for the Jost coefficient $b(\lambda, t)$, and assuming that the soliton is fast, we obtain for the emission amplitude the equation

$$\frac{\mathrm{d}b}{\mathrm{d}t} = (\mathrm{i}\pi/8)Ak^2(\lambda^2+\eta^2)^{-1}\operatorname{sech}[\pi(k+2\lambda)/4\eta]\mathrm{e}^{\mathrm{i}[k-4(\lambda^2+\eta^2)]t}. \tag{3.32}$$

Using the adiabatic variation of the perturbation, and believing that perturbation is absent at the infinity, i.e. $t \to -\infty$, we obtain the following expression for the Jost coefficient $b(t)$:

$$b^*(t) = -\frac{\pi}{8}Ak^2(\lambda^2+\eta^2)^{-1}\operatorname{sech}\left[\frac{\pi(k+2\lambda)}{4\eta}\right]\{[k-4(\lambda^2+\eta^2)]+\mathrm{i}\alpha\}^{-1}$$
$$\times \exp(-\mathrm{i}\{[k-4(\lambda^2+\eta^2)]t+\varphi_0\}). \tag{3.33}$$

Using the relation

$$\frac{\mathrm{d}}{\mathrm{d}t}N(t,\lambda) = \frac{2}{\pi}\operatorname{Re}\left\{b^*\frac{\mathrm{d}b}{\mathrm{d}t}\right\} \tag{3.34}$$

for the spectral density of the quantum number N, one can obtain the result

$$\frac{\mathrm{d}N(t,\lambda)}{\mathrm{d}t} = \frac{1}{\pi}(\pi Ak)^2\operatorname{sech}^2\frac{\pi(k+2\lambda)}{4\eta}\,\delta\left[\lambda^2-\left(\frac{k}{4}-\eta^2\right)\right]. \tag{3.35}$$

The total plasmon number emission rate is given by (3.34)

$$\frac{\mathrm{d}N}{\mathrm{d}t} = \int_{-\infty}^{\infty}\mathrm{d}\lambda\frac{\mathrm{d}}{\mathrm{d}\lambda}N(\lambda) = \left(\frac{\mathrm{d}N}{\mathrm{d}t}\right)_+ + \left(\frac{\mathrm{d}N}{\mathrm{d}t}\right)_-. \tag{3.36}$$

Here

$$\left(\frac{\mathrm{d}N}{\mathrm{d}t}\right)_\pm = \tfrac{1}{8}(\pi Ak)^2(k-4\eta^2)^{-1/2}\operatorname{sech}^2[(\pi/4\eta)\surd(k-4\eta^2)\mp k)].$$

From this expression it follows that the emission is concentrated near the points

$$2\lambda_\pm = \pm\surd(k-4\eta^2).$$

It is clear that two groups of waves exist: those emitted to the right and those emitted to the left. The group velocity of emitted waves is

$$v_{\mathrm{gr}} = -4\lambda. \tag{3.37}$$

The final example is the emission from a KdV soliton in the periodically inhomogeneous medium. The corresponding perturbed KdV equation is

$$u_t - 6uu_x + u_{xxx} = (\varepsilon_0/2\lambda)\sin[(x+t)/\lambda]u. \tag{3.38}$$

This equation arises in the investigation of ion-sound soliton evolution in inhomogeneous plasma [3.24].

The emitted field is described by the Jost coefficient $b(k,t)$. At the first order of perturbation theory, this coefficient can be calculated from the equation

$$\frac{\mathrm{d}b(k,t)}{\mathrm{d}t} = 8\mathrm{i}k^3 b(k,t) - \frac{\mathrm{i}\varepsilon\exp(-2\mathrm{i}k\xi)}{2k\kappa(k^2+\kappa^2)}\int_{-\infty}^{\infty}\mathrm{d}z\, R[u_s(z)](k-\mathrm{i}\kappa\tanh z)^2\exp(-2\mathrm{i}kz/\kappa), \tag{3.39}$$

where the perturbation $\varepsilon R = (\varepsilon_0/2\lambda)\sin[(x+t)/\lambda]$, and $u_s(z)$ is the one-soliton solution of the KdV equation $u_s(z) = -2\kappa^2\,\mathrm{sech}^2\, z$.

We introduce the quantity $J(k) = \mathrm{Re}[b(k,t)\mathrm{d}b^*(k,t)/\mathrm{d}t]$. Radiative contribution of many physical values, like the mass M, the field momentum Π and the energy, E can be expressed through $J(k)$. For example (see Appendix 3)

$$M = \int_{-\infty}^{\infty} u(x)\mathrm{d}x = M_{\mathrm{d}} + M_{\mathrm{c}} \tag{3.40}$$

$$E = -\int_{-\infty}^{\infty}\left(u^3 + \frac{1}{2}u_x^2\right)\mathrm{d}x = E_{\mathrm{d}} + E_{\mathrm{c}} = \frac{32}{\pi}\kappa^5 + \frac{32}{\pi}\int_{-\infty}^{\infty} k^4|b|^2\,\mathrm{d}k$$

Then we have

$$\frac{\mathrm{d}E_{\mathrm{c}}}{\mathrm{d}t} = \frac{32}{\pi}\int_{-\infty}^{\infty} k^4 J(k)\,\mathrm{d}k. \tag{3.41}$$

Substituting the expressions for the perturbation and (3.40) into (3.39) we obtain

$$J = \left(\frac{\pi\varepsilon_0}{\lambda^2 k}\right)^2 \frac{1}{\sinh^2\left(\dfrac{\pi}{\sqrt{v_{\mathrm{B}}\lambda}}\right)}\left[\delta\left(2v_s k - \frac{1}{\lambda}\right) + \delta\left(2v_{\mathrm{B}}k + \frac{1}{\lambda}\right)\right] \tag{3.42}$$

For $\mathrm{d}M_\mathrm{c}/\mathrm{d}t$ we have

$$\frac{\mathrm{d}M_\mathrm{c}}{\mathrm{d}t}=\frac{4}{\pi}\int_{-\infty}^{\infty}J(k)\,\mathrm{d}k=2\left(\frac{2\pi\varepsilon_0 v_\mathrm{s}}{\lambda}\right)^2\frac{1}{\sinh^2\left(\dfrac{\pi}{\sqrt{v_\mathrm{B}}\lambda}\right)}.$$

3.3 EMISSION IN A NONSTATIONARY MEDIUM

As shown in the previous section, there is a transient emission of linear waves by solitons passing through the interface between different media, resembling the emission by charged particles. At the same time it is known that nonstationarity of the medium leads to a transient scattering and the emission of electromagnetic waves by particles [3.8], [3.9]. So it is of interest to study the solitonic analog of the transient scattering of charged particles in a nonstationary medium.

Here we consider the problem of emission by the SG solitons in a medium with specific models of nonstationarity [3.10].

The wave equation describing SG soliton dynamics in a nonstationary medium takes the form

$$u_{tt}-u_{xx}+(1+\varepsilon(t))\sin u=0. \tag{3.43}$$

This problem arises, for example, in the study of magnetic soliton dynamics in planar quasi-one-dimensional magnetic media under the joint action of a constant and a variable magnet field, applied along the x- and y-axis respectively.

Let us study the problem of the emitted field and the energy calculation, for various models of the nonstationary medium. Let us consider the case of weak perturbations, i.e. $\varepsilon \ll 1$. Then we will find the solution in the form

$$u=u_\mathrm{s}+u_1+u_2+\cdots, \tag{3.44}$$

where u_s is the single-soliton solution (see Section 1.3.4). Let us use a references frame, moving with a soliton, with coordinates

$$x'=\frac{x-vt}{\gamma},\quad t'=\frac{t-vx}{\gamma},\quad \gamma=\sqrt{(1-v^2)}. \tag{3.45}$$

We have

$$u_{t't'}-u_{x'x'}+\sin u=\varepsilon[(vx'+t')/\gamma]\sin u. \tag{3.46}$$

Substituting (3.45) into (3.46) and ignoring the terms of order ε^2, one obtains an equation for the correction w_1,

$$u_{1t't'} - u_{1x'x'} + \cos u_s(x')u_1 = -\varepsilon[vx' + t')/\gamma] \sin u_s. \tag{3.47}$$

Let us consider the Fourier components $\tilde{u}_1$ in (3.47), for which

$$\tilde{u}_{1t't'}(\omega) - \omega^2\tilde{u}_1 + \cos u_s(x')\tilde{u}_1(\omega) = \varepsilon(\omega_1) \exp(-i\omega vx') \sin u_s(x'), \qquad \omega_1 = \gamma\omega. \tag{3.48}$$

Then for $\tilde{u}_1$ one has

$$u_1(x', \omega) = 2\varepsilon(\omega_1) \int_{-\infty}^{\infty} \exp(-i\omega vx')(\sinh x'/\cosh^2 x')G_\omega(x, x')\, dx', \tag{3.49}$$

where $G_\omega(x, x')$ is the Green function of the linear operator

$$L = \partial^2/\partial z^2 - \omega^2 + \cos u_s(x).$$

This function has been found by Mkrtchyan and Schmidt [3.4] and is equal to

$$G_\omega(x, x') = -\frac{1}{2q(1-q^2)} \begin{cases} e^{-q(x-x')}(\tanh x + q)(\tanh x' - q), x > x' \\ e^{-q(x'-x)}(\tanh x' + q)(\tanh x - q), x < x' \end{cases} \tag{3.50}$$

Further we will take interest in the emission field far from a soliton, so in (3.49) the integration is carried out between the limits $|z'| = 1$, under the condition $|z| \gg 1$, to lead to the expression

$$u_1(x, \omega) = \frac{\varepsilon(\omega_1)v \exp(-i\omega vx)\omega\varphi}{i\omega \cosh x} + \frac{\pi^2\gamma\varepsilon(\omega_1)}{2ip_0}\left[\frac{\theta(x)\exp(ip_0x)}{\cosh[\pi(\omega v + p_0)/2]}(\tanh x - ip_0) + \frac{\theta(-x)\exp(-ip_0x)(\tanh x + ip_0)}{\sinh[\pi(\omega v - p_0)/2]}\right]. \tag{3.51}$$

The first term in this expression is a field, localized on the soliton, and the last two correspond to the forward and backward emission.

Let us consider concrete models of the nonstationary medium:

(a) A discontinuous change of characteristic frequency in time:

$$\varepsilon(t) = \varepsilon_0 \theta(t - t_0), \qquad \theta(t) = \begin{cases} 1, & t > 0 \\ 0, & t < 0. \end{cases}$$

Performing the Fourier transformation of (3.3.9), one obtains the emitted field:

$$u_1^{\mp}(x, t) = \frac{\varepsilon\gamma^2 \theta(\pm x)}{2} \left\{ -\tanh x \int_{-\infty}^{x \mp t \pm \gamma t_0} \mathrm{d}s \tan^{-1} \sinh (s/(s + v)) + \tan^{-1} \sinh \left(\frac{x \mp t \pm \gamma t_0}{1 \pm v^2} \right) \right\}. \tag{3.52}$$

As can thus be seen, for the step change of characteristic frequency of the medium in time, the soliton emits forward and backward waves. This is in agreement with the charged-particle analogy for solitons. Besides, as follows from the form of the first term in equation (3.51), there is also a wave moving with the soliton and causing the distortion of its profile:

$$u_1^{(1)}(x, t) = \varepsilon_0 v(t - t_0) \cosh^{-1}[(x - vt)/\gamma] \theta(t - t_0).$$

(b) the model of slowly change of the medium parameters,

$$\varepsilon(t) = \varepsilon_0 \tanh (t/T), \qquad \varepsilon(\omega) = \mathrm{i}\varepsilon_0 T / \sinh (\pi\omega T/2).$$

Calculating the integrals in (3.50) for the emission field, one finds

$$u_1^{(\pm)}(x, t) = \pi\gamma^3 \varepsilon_0 T \theta(\pm x) \sum_{m=1}^{\infty} (-1)^m \frac{\exp\{-2m|x \mp t|/T\gamma\}}{\cos [\pi(1 \pm v)m/T\gamma]} \times \left[\frac{\tanh (x)}{m} - \frac{1}{T\gamma} \right] - \frac{\exp\{-2m|x \mp t|/(1 \pm v)\}}{\sin [\pi T\gamma(2m - 1)/2(1 \pm v)]} \times \left[\frac{\tanh (x)}{2m - 1} - \frac{1}{1 + v} \right]. \tag{3.53}$$

The sign $(+)$ corresponds to waves emitted forwards by the soliton, the sign $(-)$ —to the waves emitted backwards.

(c) the model for a smooth pulse in $\varepsilon(t)$,

$$\varepsilon = \frac{\varepsilon_0}{\cosh (t/T)}, \qquad \varepsilon(\omega) = \frac{\varepsilon_0 T}{\cosh (\pi T\omega/2)}.$$

The emitted field is equal to

$$u_1^{(\pm)}(x,t) = \pi\gamma^3\varepsilon_0 T(\pm 1)\theta(\pm x)\left\{\frac{\exp\{-(2m+1)|x \mp t|/(1+v)\}}{\cos[\pi T(2m+1)/2(1\pm v)]}\right.$$

$$\times\left|\frac{\tanh(x)\mathrm{sgn}(x\pm t)}{2m+1}+\frac{1}{1\pm v}\right|$$

$$\left.+\frac{\exp[-(2m+1)|x\pm t|/T\gamma]}{\cos[\pi(1\pm v)(2m+1)/2T\gamma]}\left[\frac{\tanh(x)\mathrm{sgn}(x\mp t)}{2m+1}+\frac{1}{T\gamma}\right]\right\}. \tag{3.54}$$

Again there are two groups of waves, emitted forwards and backwards by the soliton.

(d) a model for the case of periodically frequency modulation:

$$\varepsilon(t) = \varepsilon_0 \sin(\Omega_0 t), \qquad \varepsilon(\omega) = \frac{\varepsilon_0}{2\mathrm{i}}[\delta(\Omega_0-\omega)-\delta(\Omega_0+\omega)].$$

The emitted field here takes the form

$$u_1^{(\pm)}(x,t) = -\frac{\pi\gamma^2}{4} - \theta(\pm x)\left\{\frac{\tanh(x)}{k_0} \quad \frac{\cos(k_0 x \mp \Omega_0 t/\gamma)}{\cosh\left[\frac{\pi}{2}(\Omega_0/\gamma \pm k_0)\right]}\right.$$

$$-\frac{\cos(k_0 x \pm \Omega_0 t/\gamma)}{\cosh[(\pi/2)(\Omega_0/\gamma \mp k_0)]}+\frac{\sin(k_0 x \mp \Omega_0 t/\gamma)}{\cosh[(\pi/2)(\Omega_0/\gamma \pm k_0)]}$$

$$\left.-\frac{\sin(k_0 x \mp \Omega_0 t/\gamma)}{\cosh[(\pi/2)(\Omega_0/\gamma \mp k_0)]}\right\}, \qquad k_0 = \sqrt{(\Omega_0^2/\gamma^2 - 1)}. \tag{3.55}$$

The condition for wave emission in dimensional variables takes the form

$$\Omega_0 > \gamma\omega_0.$$

In the case of the planar ferromagnetic $CsNiF_3$ at a frequency $\Omega_0 >$ 100 MHz, $v > 0.6$, one observes the emission of spin waves by solitons. Thus, for periodic modulation $\varepsilon(t)$, two waves, running in opposite directions, are created. The velocity threshold, above which the emission arises, takes the form

$$v_{\mathrm{thr.}} = \sqrt{(1-\Omega_0^2/\omega_0^2)}. \tag{3.56}$$

The field u_1 diverges as $v \to v_{\text{thr}}$. In this region our calculation is not reasonable, since u_1 becomes greater than 1.

Let us calculate the emitted energy density in the nonstationary medium. Let u_1 be the emitted field. Then the energy density is

$$E = \tfrac{1}{2}\{u_{1t}^2 + u_{1x}^2 + u_1^2\}, \tag{3.57}$$

and the energy flux $J = u_{1x}u_{1t}$ satisfies the relationship

$$E_t - J_x = 0. \tag{3.58}$$

Let us designate the forward emitted energy by $\bar{E}^{(+)}$ and that emitted backwards by $\bar{E}^{(-)}$,

$$\bar{E}^{(+)} = \int_0^{\infty} E(x, t)\,\mathrm{d}x, \qquad \bar{E}^{(-)} = \int_{-\infty}^{0} E(x, t)\,\mathrm{d}x. \tag{3.59}$$

After long transformations to the spectral density of the emitted energy one obtains the following equation:

$$\frac{\mathrm{d}\bar{E}^{(\pm)}}{\mathrm{d}\omega} = \frac{\pi^3\gamma^4}{2}\,\frac{\varepsilon^*(\omega)\varepsilon(\omega)\omega^2\,\mathrm{sgn}\,\omega}{p_0(\omega \mp vp_0)\cosh^2[\pi(\omega v \pm p_0)/2]}. \tag{3.60}$$

Let us analyze the formula (3.60) for case (b). We will consider the case of small T. Then for the spectral energy density of emission there is the following expansion:

$$\bar{E}_{\omega}^{(\pm)} = \bar{E}_{\omega 0}^{(\pm)}(1 + \tfrac{1}{24}\pi^2 T^2\gamma^2\omega^2), \tag{3.61}$$

where $\bar{E}_{\omega 0}^{(\pm)}$ is the intensity of transient radiation under the discontinuous change of medium properties in case (a). In the case of high frequency one obtains

$$\bar{E}_{\omega}^{(\pm)} \approx 2\pi^3\gamma^6\varepsilon_0^2 T^2\omega\exp[-\pi\omega(T\gamma + v)], \tag{3.62}$$

i.e. the intensity of transient emission from a smeared nonstationary layer is exponentially small. The exponential decrease of the spectrum is due to the fact that the time-dependence $\omega(t)$ is a smooth continuous function with derivatives of all orders. This fact is in agreement with the results for transient charge emission in nonstationary medium and represents another facet of the analogy between solitons and particles, as pointed out by many authors. In the nonuniform plasma such an analogy was noted in the articles [3.11], [3.12].

We will consider also the emission from NLS solitons in a nonstationary medium. The model is described by the perturbed NLS equation with a potential $n(x, t)$ [3.14],

$$iq_t + q_{xx} + 2|q|^2 q = -n(x, t)q,$$

$$n(x, t) = \sum_l \delta(t - l\tau) n(x). \tag{3.63}$$

Applying the perturbation theory for NLS solitons (see Appendix 1), we obtain equations for the soliton parameters η, ξ, prior to the action of the single δ-potential, and after the action of the potential, $\bar{\eta}$, $\bar{\xi}$:

$$\begin{aligned} \bar{\eta} &= \eta, \\ \bar{\xi} &= \xi - \tfrac{1}{2} n_x(x_0). \end{aligned} \tag{3.64}$$

Taking into account that $v = -4\xi$, we have

$$\begin{aligned} \bar{v} &= v + 2n_x(x_0), \\ \bar{x} &= x_0 + \tau\bar{v}. \end{aligned} \tag{3.65}$$

If $n_0 = \cos\gamma x$, then (3.65) is the well-known *standard map*, having chaotic properties [3.12]. The problem concerning the chaotic behavior of solitons will be considered in Chapter 5. Here we will solve the problem of soliton emission under temporal periodic δ-like modulation of the medium [3.11]. First, we note that a small parameter of this problem is $|n_{xx}|$. That is clear from the expansion

$$n(x) = n(x_0) + n'(x_0)(x - x_0) + \tfrac{1}{2} n_0''(x - x)^2 + \cdots.$$

As was shown in Section 2.2, the influence of the first and second terms does not lead to radiation: the only consequence is that the soliton parameters are renormalized. To calculate the emitted power it is necessary to find the Jost coefficient $b(k, t)$. To the first order in the small-parameter expansion, we have

$$|b|^2 = 1 - \left| \int_{-\infty}^{\infty} [\delta u(\varphi^{(2)})^2 + \delta u^*(\varphi^{(1)})^2]\,dx \right|^2, \tag{3.66}$$

where $\varphi^{(1,2)}$ are the Jost function components for a single soliton (Appendix 1). Using expression (A.34) for $E(k)$, we obtain for the emitted power

$$E(k) = \frac{1}{\pi}\left|\int_{-\infty}^{\infty} \frac{2\eta}{(\lambda-\lambda_1^*)^2} \frac{e^{i(kx-\theta)}}{\cosh z}\right.$$

$$\left[\left(\exp\left(i\frac{n_{xx}(x_0)}{2}(x-x_0)^2\right)-1\right)(\lambda-\xi-i\eta\tanh z)^2\right.$$

$$\left.\left. -\left(\exp\left(-\frac{n_{xx}(x_0)}{2}(x-x_0)^2\right)-1\right)\frac{\eta^2}{\cosh^2 z}\right]dx\right|^2. \quad (3.67)$$

After long integrations we have the result

$$E(k) = \frac{\pi|n_{xx}(x_0)|^2}{16|\lambda-\lambda_1|^4} \frac{1}{\cosh[(k-\bar{\xi})/2\eta\pi]}. \quad (3.68)$$

It is seen that the radiation line is exponentially narrow and that there is a Doppler shift associated with the soliton motion. The radiation effect leads to soliton damping, as is obvious from the relation

$$N = \int_{-\infty}^{\infty} |q|^2\,dx = \int_{-\infty}^{\infty} E(k)\,dk + 4\eta. \quad (3.69)$$

Substituting (3.68) into (3.69), we find the law of soliton radiation damping:

$$\bar{\eta} = \eta - \pi(n_{xx}(x_0))^2 M_0/(64\eta^3), \quad (3.70)$$

where

$$M_0 = \int_{-\infty}^{\infty} \frac{1}{\cosh^2[t(t^2+1)/2\pi]}\,dt.$$

Damping of the soliton amplitude, despite the fact that the perturbation is conservative, has a simple explanation. The total quantum number is conserved, i.e. the system 'soliton + radiation' is conserved. Under the action of the medium inhomogeneities, the soliton is damped by radiation. In an infinite medium the radiation propagates to infinity, and is not absorbed by solitons. The calculation shows that the radiation leads to renormalization of the force acting on solitons (this may be termed 'radiative viscosity').

3.4 RADIATION OF SOLITONS IN A MEDIUM WITH MOVING INHOMOGENEITIES

Let us now consider the emission of waves by solitons scattered by a moving localized inhomogeneity. A typical example of this problem is in the dynamics of coupled magnetic and acoustic solitons in an anharmonic magnetic crystal. The two kinds of solitons move with different velocities. As a result, an acoustic soliton induces the above-mentioned moving inhomogeneity of magnetic medium with respect to magnetic solitons.

We will consider magnetic soliton dynamics in anharmonic planar ferromagnetic crystals. As was shown in Ref. [3.14], the phonon deceleration of magnetic solitons in a number of rare-earth orthoferrites under the condition of supersonic soliton motion is very important. In this case, when the velocity is close to the sound velocity it is necessary to take into account the influence of lattice anharmonicity. The investigation of magnetic soliton dynamics in linear magnetic polymers [3.15] is another important application. Since a polymer represents a rather easily deformable system, the need arises to take anharmonicity effects into account.

Let us consider the case of planar ferromagnetics [3.16], [3.17]. The total Hamiltonian for the spin–phonon system in the long-wave limit is chosen in a form including, in accordance with [3.18], the anharmonic addition to the phonon part of the Hamiltonian,

$$
\begin{aligned}
H_{\mathrm{tot}} &= H_{\mathrm{ph}} + H_{\mathrm{s}} + H_{\mathrm{int}}, \\
H_{\mathrm{ph}} &= \frac{1}{2}\sum_i \frac{p_i^2}{2m} + \frac{\kappa}{2}(u_{i+1} - u_i)^2 + \frac{\beta}{4}(u_{i+1} - u_i)^4, \\
H_{\mathrm{s}} &= \sum_i (-I\bar{S}_i\bar{S}_{i+1} + A(S_i^z)^2 - 2g\mu_{\mathrm{B}}H^x S_i^x, \\
H_{\mathrm{int}} &= \alpha \sum_i (u_{i+1} - u_i)\bar{S}_i\bar{S}_{i+1},
\end{aligned}
\tag{3.71}
$$

where u_i are the lattice displacements, a is the lattice constant, $\bar{S}$ is the classical spin vector, I is the exchange integral, A is the anisotropy constant ($I, A > 0$), H^x is the external constant magnetic field, p_i is the momentum, α is the spin–phonon coupling constant, m is the mass of a lattice atom, κ is the elastic constant, and β is the anharmonicity coefficient. We consider the long-wave limit, when the scales of wave processes are greater than the lattice constant a. We have

$$\bar{S}_i = S(z), \qquad \bar{S}_{i+1} = S(z) + a\frac{\partial S}{\partial z} + \frac{a^2}{2}\frac{\partial^2 S}{\partial z^2} + \cdots,$$
$$u_i = u(z), \qquad u_{i+1} = u(z) + a\frac{\partial u}{\partial z} + \frac{a^2}{2}\frac{\partial^2 u}{\partial z^2} + \cdots. \tag{3.72}$$

Substituting (3.72) into (3.71), we find for the Hamiltonian density

$$H = \frac{Ia^2}{2}\left(\frac{\partial S}{\partial z}\right)^2 + A(S^z)^2 - 2\mu_B g H^x S^x + \frac{p^2}{2m} + \frac{\kappa a^2}{2}\left(\frac{\partial u}{\partial z}\right)^2 + \frac{\kappa a^3}{2}\frac{\partial u}{\partial z}\frac{\partial^2 u}{\partial z^2} + \frac{\kappa a^4}{12}\frac{\partial u}{\partial z}\frac{\partial^3 u}{\partial z^3} + \frac{\beta a^4}{24}\left(\frac{\partial u}{\partial z}\right)^4 + \frac{\alpha a^3}{2}\frac{\partial u}{\partial z}\left(\frac{\partial S}{\partial z}\right)^2. \tag{3.73}$$

For the case of very small temperature, the classic spin modulus is not changed, i.e. possible changes are reduced to those of direction. For this reason it is useful to introduce the angle variables

$$\bar{S} = S_0\{\sin\theta\cos\varphi, \sin\theta\sin\varphi, \cos\theta\}, \qquad \bar{S}^2 = S_0^2.$$

Then the system of equations for $\varphi(x,t)$ and $u(x,t)$ takes the form

$$\varphi_{tt} - c_0^2\varphi_{zz} + \omega_0^2\sin\varphi = \alpha c_0^2 a(u_z\varphi_z)_z/I, \tag{3.74}$$

$$u_{tt} - v_a^2\left(u_{zz} + \frac{a^2}{12}u_{zzzz}\right) - \frac{3\beta a^4}{m}(u_z)^2 u_{zz} = \alpha S_0^2 a^3(\sin^2\theta\,\varphi_z^2 + \theta_z^2)_z/2m. \tag{3.75}$$

$$\cos\theta = -(1/2AS_0)\varphi_t, \tag{3.76}$$

where

$$\omega_0^2 = 2g\mu_B H^x AS, \quad c_0^2 = 2AS_0^2a^2I, \quad v^2 = \kappa a^2/m.$$

The complete solution of this system is difficult to find. Some particular cases may be considered with the aid of perturbation theory. We will consider the case when one soliton is cited in the magnetic system, and one supersound acoustic soliton in the acoustic subsystem. Let us use the variable $\zeta = x - vt$. We ignore the term $\sim \alpha S^2 a^3$ in the first approximation in the right-hand side of the equation (3.77). On integrating the equation (3.78), one can obtain that [3.18]

$$\varphi = 4\tan^{-1}\exp\left(\pm\frac{\omega_0(z - v_m t)}{c_0\sqrt{(1 - v_m^2/c_0^2)}}\right),$$
$$u_s = \pm\sqrt{(2(v_u^2/v_a^2 - 1)\kappa/\beta)}\,\mathrm{sech}\, q(z - v_u t), \tag{3.77}$$
$$q^2 = 12(v_u^2/v_a^2 - 1)/a^2, \qquad v_u > v_a.$$

Here v_u is the acoustic soliton velocity, v_m is the magnetic soliton velocity, v_a is the sound velocity in the chain, c_0 is the velocity of spin waves. We use below the dimensionless variables

$$\beta_m = v_m/c_0,\ \beta_u = v_u/c_0,\ \Delta\beta = \beta_u - \beta_m,$$
$$\Delta\gamma = 1 - \beta_m\beta_u,\ t' = \omega_0 t,\ z' = \omega_0 z/c_0. \tag{3.78}$$

To investigate the influence of the elastic subsystem on the magnetic, one we solve equation (3.74). We assume that the spin–phonon interaction is small and the solution of equation (3.74) is found in the form

$$\varphi = \varphi_s(z', t') + \varphi_1 + \varphi_2 + \cdots, \qquad \varphi_1 \sim \alpha, \tag{3.79}$$

Substituting (3.79) into (3.74) one obtains the equation for the correction in dimensionless variables:

$$\varphi_{1t't'} - \varphi_{1z'z'} + \cos\varphi_s\varphi_1 = \varepsilon(u_{sz'}\varphi_{sz'})_{z'}, \qquad \varepsilon = \alpha a\omega_0/Ic_0, \tag{3.80}$$

Because the coefficients of this equation depend on the coordinate and time in the combination $(z' - \beta_m t)$, we now take into consideration a reference frame moving with the soliton:

$$x = \gamma_m(z' - \beta_m t'), \quad \tau = \gamma_m(t' - \beta_m z'), \quad \gamma_m = \sqrt{(1 - \beta_m^2)}. \tag{3.81}$$

Performing the Fourier transformation

$$\varphi_1(x', t') = \int_{-\infty}^{\infty} \exp(-i\omega\tau)\bar{\varphi}_1(x, \omega)\,d\omega, \tag{3.82}$$

one obtains

$$(-\omega^2 - \partial^2/\partial x^2 + \cos\varphi_s(x))\bar{\varphi}_1(x, \omega)$$
$$= \frac{\pi\alpha}{I}\left(\frac{\kappa}{6\beta}\right)^{1/2}\frac{(\Delta\gamma/\Delta\beta + \beta_m}{\Delta\beta\sqrt{(1 - \beta_m^2)}}\mathrm{sech}\left(\frac{\pi\omega}{2q\Delta\beta\gamma_m}\right)$$
$$\times\left[i\omega\frac{\partial^2\varphi_s}{\partial x^2} + (\Delta\gamma/\Delta\beta + \beta_m)\omega^2\frac{\partial\varphi_s}{\partial x}\right]e^{i(\Delta\gamma/\Delta\beta)\omega x}. \tag{3.83}$$

Hence

$$\tilde{\varphi}_1(x, \omega) = \varepsilon \int_{-\infty}^{\infty} \mathrm{d}x' G_\omega(x, x') R(x', \omega) \mathrm{d}x' \tag{3.84}$$

where G_ω is the Green function of the linear operator defined in Section 3.3, and $R(x, \omega)$ is the expression in the right-hand side of equation (3.83). Substituting into (3.84) the corresponding expressions and calculating integrals, one obtains for the correction

$$\begin{aligned} \bar{\varphi}_1(x, t) = {} & \mathrm{i}\frac{\pi\alpha}{\mathrm{I}}\left(\frac{\kappa}{6\beta}\right)^{1/2} \gamma_{\mathrm{m}} \frac{(\Delta\gamma/\Delta\beta + \beta_{\mathrm{m}})\omega(3\Delta\gamma^2/\Delta\beta^2) + \beta_{\mathrm{m}}\Delta\gamma/\Delta\beta + 1)}{\Delta\beta(1 \mp \mathrm{i}\sqrt{(\omega^2 - 1)})} \\ & \times \operatorname{sech}\left(\frac{\pi\omega}{2\Delta\beta\gamma_{\mathrm{m}}}\right) \operatorname{sech}\left\{\frac{\pi}{2}\left[\frac{\Delta\gamma}{\Delta\beta}\omega \pm \sqrt{(\omega^2 - 1)}\right]\right\} \\ & \times \exp(\mp \mathrm{i}\sqrt{(\omega^2 - 1)}x). \end{aligned} \tag{3.85}$$

This equation describes spin wave emission by the magnetic soliton during its interaction with the elastic subsystem.

Let us calculate the losses of soliton energy for spin wave emission. In the linear approximation the emitted field satisfies the Klein–Gordon equation. This allows us to write down the equation defining the energy density:

$$W(\omega) = \tfrac{1}{2}(k^2 + \omega^2 + 1)|\bar{\varphi}_1(x, \omega)|^2. \tag{3.86}$$

Therefore, the emitted power is equal to

$$\begin{aligned} W_\omega = {} & \left(\frac{\pi\alpha}{\mathrm{I}}\right)^2 \left(\frac{\kappa}{6\beta}\right) \gamma_{\mathrm{m}}^2 \left(\frac{\Delta\gamma}{\Delta\beta} + \beta_{\mathrm{m}}\right)^2 \left(3\frac{\Delta\gamma^2}{\Delta\beta^2} + \beta_{\mathrm{m}}\frac{\Delta\gamma}{\Delta\beta} + 1\right) \frac{\omega^4}{\Delta\beta} \\ & \times \operatorname{sech}^2\left(\frac{\pi\omega}{2q\Delta\beta\gamma_{\mathrm{m}}}\right) \operatorname{sech}^2\left\{\frac{\pi}{2}\left[\frac{\Delta\gamma}{\Delta\beta}\omega + \sqrt{(\omega^2 - 1)}\right]\right\}. \end{aligned} \tag{3.87}$$

It is clear from this expression that while crossing the acoustic soliton the magnetic one emits linear spin waves with Doppler shifted frequencies in both directions. The radiation frequency starts from $\omega = \omega_0$, and decreases monotonically when the frequency grows.

It is necessary to note that there is coupling between the acoustic perturbation amplitude and the magnetic soliton velocity. Applying the results of Ref. [3.18], we can see that if we require coincidence of the velocity along the x-axis of the acoustic and magnetic soliton velocities, then we have the equation

$$v_{\mathrm{a}}^2 + (\alpha^2 B u_0/12)^2 = v^2 \leqslant c_0^2 = 2AIS^2a^2, \tag{3.88}$$

i.e. the bound state of the magnetic and acoustic soliton is possible only for parameter values defined by expression (3.88). The interesting new phenomenon which appears is the Cherenkov emission of sound by the magnetic soliton moving with supersonic velocity. This problem was intensively studied in Ref. [3.23]. Influence of ferroelectric subsystem has been studied in Ref. [3.28].

3.5 EMISSION OF WAVES BY SOLITARY WAVES IN NONINTEGRABLE SYSTEMS

We will consider here the processes of wave emission by solitary waves in nonintegrable systems. Following the article [3.21], we study the emission by Langmuir solitons during the propagation in a medium with slow gradient of plasma concentration. We will produce a description of this phenomenon on the basis of the Zakharov system base (1.59–60).

Let us find first the acceleration of a langmuir soliton. The soliton coordinate center is defined by the relation

$$x_0 = \int_{-\infty}^{\infty} x|E|^2 \mathrm{d}x \left(\int E^2 \mathrm{d}x \right)^{-1}. \tag{3.89}$$

Differentiating (3.89) with refers to time and integrating by parts, we obtain the result

$$\int_{-\infty}^{\infty} E^2 \mathrm{d}x \frac{\mathrm{d}^2 x_0}{\mathrm{d}t^2} = \int_{-\infty}^{\infty} n \frac{\partial |E|^2}{\partial x} \mathrm{d}x = -\int_{-\infty}^{\infty} |E|^2 \frac{\partial n}{\partial x} \mathrm{d}x. \tag{3.90}$$

The perturbation of the plasma concentration may be divided into three parts: the solitonic part, the externally imposed part, and the part connected with sound emission. In the first approximation choose the soliton moving with a *given* acceleration. Then in the equation

$$n_{tt} - n_{xx} = (|E|^2)_{xx}, \tag{3.91}$$

the right-hand side is known. From this may be found the sound field emitted by solitons:

$$\begin{aligned} n_{\mathrm{rad}}(x, t) = \frac{E_0^2 v}{k_0} \Bigg\{ & \tanh k_0 (x - k \left(1 - \int_0^t v \, \mathrm{d}\tau \right) \left[\frac{1}{(1+v)^3} + \frac{1}{(1-v)^3} \right] \\ & + \frac{k_0 (x - \int_0^t v \, \mathrm{d}\tau)}{\cosh^2 k_0 (x - \int_0^t v \, \mathrm{d}\tau)} \frac{2v}{1 - v^2} \left[\frac{1}{1 - v^2} + \frac{1}{(1-v)^2} \right] \Bigg\}. \end{aligned} \tag{3.92}$$

Substituting the expression into the right-hand side of equation (3.89) we find the expression for the soliton acceleration:

$$\frac{\mathrm{d}v}{\mathrm{d}t} = \tfrac{3}{2} v_{T_e} \frac{\partial \ln n_0}{\partial x} - \frac{c_s^2 + 5v^2}{(c_s^2 - v^2)^4} v_{T_e}^2 c_s^4 \frac{\delta n_0}{n_0} \dot{v}. \tag{3.93}$$

The last term takes into account the sound back-reaction on the soliton and dominates at large values of the soliton amplitude, i.e. when $\delta n/n \gg m/M$. As a result of the acceleration the soliton velocity tends to the sound velocity for $t \to \infty$:

$$c_s - v \sim t^{-1/3}. \tag{3.94}$$

This equation was obtained also in Ref. [3.22].

If the soliton is moved in the field of a low-frequency sound wave then the soliton is sequentially accelerated or decelerated. The resulting mean acceleration for many periods was calculated in Ref. [3.23]. The acceleration is directed along the sound wave motion direction and is given by

$$\frac{\mathrm{d}}{\mathrm{d}t} M_{\mathrm{eff}} v = \frac{1}{4}\frac{k}{\omega} R_{\mathrm{ref}}^2 n_{\mathrm{sd}}^2 \frac{1 - kv/\omega}{1 + kv/\omega}. \tag{3.95}$$

The emission of sound by accelerated Langmuir solitons was considered also in Ref. [3.24] for some models of the inhomogeneities of the media. In the case of arbitrary velocities the soliton acceleration under changes of field momentum and of the number of quanta is described by the equation [3.21]:

$$\dot{v} = \frac{\mathrm{d}p}{\mathrm{d}t} - \frac{v_{T_e}^2}{c_s^2 - v^2} \left|\frac{\delta n_c}{n_c}\right| \left(\frac{v(c_s + 5v^2)}{c_s^2 - v^2} + 3\gamma v \right);$$

$$\gamma \equiv \frac{\mathrm{d}}{\mathrm{d}t} \ln \int_{-\infty}^{\infty} |E|^2 \, \mathrm{d}x. \tag{3.96}$$

4

DYNAMICS OF SOLITONS IN RANDOM FIELDS AND MEDIA

In this book we generally consider the problems arising when a nonlinear wave propagates in regular inhomogeneous media. In real systems there always exist fluctuations of the medium parameters, external fields and initial conditions that will result in the occurrence of *stochastic* dynamics of solitions and other nonlinear waves. Thereby, apart from the stochastization of the soliton parameters, processes of soliton-radiated random waves, stochastic soliton and breather decay, etc., might also occur. Moreover, in real systems, so-called 'dynamical chaos' of nonlinear waves connected with the strong instability waves under periodic perturbations (see Chapter 5), will be frequently realized, along with stochasticity induced by random sources, and one should distinguish between these mechanisms. In the case of nonlinear systems with a finite number of degrees of freedom, relevant analysis shows that dynamical chaos is a more rapid process, and comes into action essentially sooner than normal stochasticity [4.1].

In this chapter we will consider first of all the history of the problem, and the quasilinear methods, describing the evolution of nonlinear waves under random perturbations, and we outline effective functional methods. We will also compare various methods applied in the theory of random nonlinear waves, namely, mean field methods, the adiabatic approximation and the statistical Born approximation. To derive an applicability criterion, exactly solvable stochastic nonlinear waves models are used. More comprehensive review of solitons in random media is given in Refs. [4.51] and [4.52].

4.1 THE MEAN FIELD METHOD

Historically, one of the first methods developed for the description of wave evolution in randomly inhomogeneous media was the mean field method. It turned out to be rather successful in the theory of linear wave propagation (see, for example, [4.2]). As applied to the nonlinear wave theory, it was proposed in Ref. [4.3]. As a further analysis showed in the case of nonlinear wave propagation a direct application of this method leads to some serious difficulties [4.4], [4.5].

Let us first of all describe briefly the essence of the mean field method [4.6]. Consider a nonlinear wave equation

$$Lu = \varepsilon(x,t)Mu + Qu^2, \tag{4.1}$$

where Q is a deterministic linear operator, L, M are linear integro-differential operators $\varepsilon(x,t)$ is a random function with prescribed statistics. The wave field will be represented as the sum of mean $\langle u\rangle$ and scattered δu fields,

$$u = \langle u\rangle + \delta u, \quad \langle \delta u\rangle = 0. \tag{4.2}$$

Now (4.1) is substituted into (4.2) and averaged over all the realizations of ε. As a result we have a set of two equations,

$$L\langle u\rangle = \langle \varepsilon M\,\delta u\rangle + Q\langle u\rangle^2 + Q\langle \delta u^2\rangle, \tag{4.3}$$

$$L\delta u = \varepsilon M\langle u\rangle + \varepsilon M\,\delta u - \langle \varepsilon M\,\delta u\rangle + Q(\delta u^2 - \langle \delta u^2\rangle) + 2Q\langle u\rangle. \tag{4.4}$$

For $\varepsilon \ll 1$ δu is readily derived to be

$$\delta u = L^{-1}\varepsilon M\langle u\rangle, \tag{4.5}$$

and the equation for the mean field has the form

$$L\langle u\rangle = N\langle u\rangle + Q\langle u\rangle^2. \tag{4.6}$$

This equation is closed with respect to $\langle u\rangle$, and so solve it one can already apply methods of the nonlinear wave theory.

Let us illustrate the application of the mean field method with the examples of stochastically perturbed KdV and NLS equations. To begin with, we will consider the KdV equation in the form

$$u_t + 6uu_x + u_{xxx} = -\alpha(t)u_x, \tag{4.7}$$

where $\alpha(t)$ models a random change of the soliton velocity, with

$$\langle \alpha \rangle = 0, \quad \langle \alpha(t)\alpha(t+\tau) = B_\alpha(\tau). \tag{4.8}$$

Let us represent the wave field in the form (4.2). Substitution of this expansion into (4.7) yields for the mean field and correction the following set of equations:

$$\langle u \rangle_t + 6\langle u \rangle \langle u \rangle_x + \langle u \rangle_{xxx} = -\langle \alpha(t)\delta u_x \rangle + \langle \delta u_x^2 \rangle, \tag{4.9}$$
$$(\delta u)_t + 6\langle u \rangle (\delta u)_x + 6\delta u \langle u \rangle_x + \delta u_{xxx} = -\alpha \langle u \rangle,$$

where we have

$$\delta u = -\int_0^t \mathrm{d}t' \int_{-\infty}^{\infty} \mathrm{d}x' G(x, x'; t, t')\alpha(t')\langle u(x', t')\rangle. \tag{4.10}$$

Here $G(x, x'; t, t')$ is the Green function of the equation

$$[\partial_t - 6\partial_x(\partial_x^2 + \langle u \rangle)]G(x, x'; t, t') = \delta(x - x')\delta(t - t').$$

For a δ-correlated random process we have

$$B_\alpha(\tau) = \sigma_\alpha^2 \delta(\tau).$$

Now we evaluate the mean value $\langle \alpha \delta u_x \rangle$. We have

$$\langle \alpha \delta u_x \rangle = -\int_0^t \mathrm{d}t' \int_{-\infty}^{\infty} \mathrm{d}x' G_{x'}(x, x'; t, t')\langle \alpha(t)\alpha(t')\rangle \langle u(x', t')\rangle_{x'}$$
$$= 2\sigma_\alpha^2 \int_{-\infty}^{\infty} \mathrm{d}x' G_{x'}(x, x'; t, t')\langle u(x', t)\rangle_{x'}, \qquad t \to t'. \tag{4.11}$$

On taking into account the fact that $G(t, t') \xrightarrow[t \to t']{} \delta(t - t')$, we find that

$$\langle \alpha \delta u_x \rangle = 2\sigma_\alpha^2 \langle u \rangle_{xx}.$$

The final equation for the mean field is then

$$\langle u \rangle_t + 6\langle u \rangle \langle u \rangle_x + \langle u \rangle_{xxx} = 2\sigma_\alpha^2 \langle u \rangle_{xx}. \tag{4.12}$$

For small σ_α^2 the following modified mean energy equation is satisfied:

$$\frac{\mathrm{d}}{\mathrm{d}t}\int_{-\infty}^{\infty} (\langle u \rangle)^2 \, \mathrm{d}x = 4\sigma_\alpha^2 \int_{-\infty}^{\infty} \langle u \rangle \langle u \rangle_{xx} \, \mathrm{d}x. \tag{4.13}$$

A solution of the form $\langle u \rangle = 2\kappa^2 \operatorname{sech}^2 \kappa(x-\xi)$ is substituted into (4.13) to yield the equation

$$\frac{\mathrm{d}\kappa}{\mathrm{d}t} = -\frac{16}{15}\sigma_\alpha^2 \kappa^3$$

for the amplitude. Thus, the mean field amplitude decreases according to the law

$$\kappa(t) = \kappa_0 (t_0 + t)^{-1/2}. \tag{4.14}$$

Let us study further a stochastic NLS equation. It is one of the most frequently appearing nonlinear wave equations in plasma physics, solid state physics, nonlinear optics, etc. The stochastic NLS equation will be taken in the form [4.7]

$$\mathrm{i}q_t + \tfrac{1}{2}q_{xx} + |q|^2 q = -\varepsilon(x,t)q, \tag{4.15}$$

appearing in the study of electromagnetic beam propagation in a randomly inhomogeneous medium, and of optical solitons in randomly nonuniform fibers. The equation for the mean field has the form

$$\mathrm{i}\langle q \rangle_t + \tfrac{1}{2}\langle q \rangle_{xx} + \langle |q|^2 q \rangle + \langle \varepsilon(x,t) q \rangle = 0. \tag{4.16}$$

According to the mean field method we put

$$\langle |q|^2 q \rangle \cong |\langle q \rangle|^2 \langle q \rangle.$$

Then it is necessary to decouple the mean $\langle \varepsilon q \rangle$. For this we will make use of the Furutzu–Novikov formula [4.8]:

$$\langle \varepsilon(x,t) q(x,t) \rangle = \int_0^t \mathrm{d}t' \int_{-\infty}^{\infty} \mathrm{d}x' \langle \varepsilon(x',t')\varepsilon(x,t) \rangle x \left\langle \frac{\delta q(x,t)}{\delta \varepsilon(x',t')} \right\rangle. \tag{4.17}$$

The causality condition requires

$$\frac{\delta q(x,t)}{\delta \varepsilon(x',t')} = 0, \qquad t' < t. \tag{4.18}$$

We will integrate (4.16) between zero and t to then calculate the variational derivative $\delta q / \delta \varepsilon(x',t')$. We thus arrive at

$$\lim \frac{\delta q(x,t)}{\delta \varepsilon(x',t')} = 2\mathrm{i}A_\varepsilon(0) \langle q(x,t) \rangle \delta(x-x'). \tag{4.19}$$

for the model of correlator $\langle \varepsilon(x',t')\varepsilon(x,t) \rangle = 2A_\varepsilon(x-x')\delta(t-t')$.

Substituting (4.19) into (4.17) and (4.16), we obtain

$$\mathrm{i}\langle q\rangle_t + \tfrac{1}{2}\langle q\rangle_{xx} + |\langle q\rangle|^2\langle q\rangle = -\mathrm{i}A_\varepsilon(0)\langle q\rangle. \tag{4.20}$$

This is the NLS equation with damping, and can be analytically analyzed for $A_\varepsilon(0) \ll 1$.

In the next chapter we will illustrate, with examples of exactly solvable nonlinear stochastic wave equations, that the mean field method is *incorrect* for such a direct separation of nonlinearity and inhomogeneity effects. Recently, in Ref. [4.9], the mean field method was modified to allow rigorous asymptotic justification. The main idea of the work involves going to a reference frame moving with a fluctuating velocity. This allow the separation of the effects of phase modulation from those of amplitude modulation, and thereby permits the study of wave profile evolution.

Consider a wave equation describing the propagation:

$$u_{tt} - (1 + \varepsilon\alpha)^2 u_{xx} = \varepsilon^2(u^2)_{xx}, \qquad \varepsilon > 0, \quad \varepsilon \ll 1. \tag{4.21}$$

Let us go to the reference frame moving with the fluctuating velocity (the case of purely temporal fluctuations), with independent variables (x', t'),

$$x' = x - \int_0^t c(t')\,\mathrm{d}t'. \tag{4.22}$$

In terms of these new variables this equation takes the form

$$u_{tt} - 2cu_{tx} - c_t u_x + [c^2 - (1 + \varepsilon\alpha)^2]u_{xx}. \tag{4.23}$$

Now it is convenient to introduce slow times $T = \varepsilon^2 t$, $T_1 = \varepsilon^3 t$, and seek the solution of (4.23) in the form of an asymptotic series

$$u(x, t, T, \ldots) = u^{(0)}(x, T, \ldots) + \varepsilon u^{(1)}(x, t; T, \ldots) + \cdots;$$
$$c(t, T, \ldots) = 1 + \varepsilon c^{(1)}(t, T, \ldots) + \cdots. \tag{4.24}$$

After some algebra (for details see [4.9]) one can obtain the equation describing slow wave evolution (script zero is omitted)

$$u_T + \frac{\sigma^2}{2}u_x + uu_x + \int_0^\infty u_{xx}(x + 2\tau, T)W(\tau)\,\mathrm{d}\tau = 0, \tag{4.25}$$

where $W(\tau) = \langle\alpha(t)\alpha(t + \tau)\rangle$ is the correlation function of the sound velocity fluctuations.

As compared with (4.12), equation (4.25) is different in that it describes a *mean form of the wave rather than a mean field*. Averaging over fast time allows not only the removal of secular terms, but corresponds most appropriately to the experimental situation—as compared with conventional averaging over the ensemble of realizations. Comparison of the results obtained with those of the a standard mean field method shows a great difference between the types of nonlinear factors. In the linear approximation, however, the two approaches coincide, in their results.

4.2 EXACTLY SOLVABLE NONLINEAR STOCHASTIC WAVE MODELS

The examination of exactly solvable nonlinear stochastic wave equation is of great interest for the study of the limits of applicability of approximate methods, in particular the mean field method. Gurvatov *et al.* [4.4] were the first to consider this exactly solvable stochastic model. Following their work, we will consider a model equation of the form

$$u_t + \alpha(t)u_x + \hat{Q}_{\ln}u + \hat{Q}_{\mathrm{nl}}u = 0, \tag{4.26}$$

where $u(x, t)$ is the wave field, $\hat{Q}_{\ln}$, $\hat{Q}_{\mathrm{nl}}$ are respectively linear and nonlinear deterministic operators, and $\alpha(t)$ is a random function of time. The following replacement of the variables,

$$X = x - \eta(t), \qquad \eta(t) = \int_0^t \alpha(t')\,\mathrm{d}t', \tag{4.27}$$

is made. In terms of the new variables, the equation turns into a deterministic one,

$$u_t + \hat{Q}_{\ln}u + \hat{Q}_{\mathrm{nl}}u = 0. \tag{4.28}$$

The solution of this equation is related to that of the original one as follows:

$$u(x, t) = U(x) - \eta(t), t). \tag{4.29}$$

Then the *mean field* is found to be

$$\langle u(x, t) \rangle = \int U(x - \eta, t) W(\eta; t)\,\mathrm{d}\eta, \tag{4.30}$$

where $W(\eta, t)$ is the probability density function of η. We will consider $\alpha(t)$ as a stationary gaussian process, with the zero mean value and correlation function $B(\tau)$: then

$$W(\eta, t) = \exp\{-\eta^2/2d(t)\}/\sqrt{(2\pi d(t))},$$

$$d(t) = 2\int_0^t (t-\tau)B(\tau)\,\mathrm{d}\tau. \tag{4.31}$$

The density W is seen to satisfy the equation

$$\frac{\partial W}{\partial d} = \frac{\partial^2 W}{2\partial\eta^2}. \tag{4.32}$$

Taking into account the equations (4.30)–(4.32), we obtain the equation for the mean field as

$$\langle u\rangle_t + \hat{Q}_{\mathrm{ln}}\langle u\rangle + \langle \hat{Q}_{\mathrm{nl}}u\rangle = A(t)\langle u\rangle_{xx};$$

$$A(t) = \int_0^t B(\tau)\,\mathrm{d}\tau,\ A(t \gg \tau_0) = A(\infty) = D. \tag{4.33}$$

The ordinary mean field method (see Section 4.1) leads to the different equation

$$\langle u\rangle_t + \hat{Q}_{\mathrm{nl}}\langle u\rangle + \hat{Q}_{\mathrm{nl}}\langle u\rangle = \int_0^t B(t-\tau)\langle u\rangle_{xx}\,\mathrm{d}\tau. \tag{4.34}$$

Comparing these equations, we find that the exact and approximate equations are distinguished by two terms: one dissipative, one nonlinear.

Let us analyze the situation for the example of the stochastic KdV equation.

$$u_t + \alpha(t)u_x + uu_x + \beta u_{xxx} = 0.$$

Consider the exact mean (4.30) for a soliton

$$u(x,t) = u_0\,\mathrm{sech}^2[\sqrt{(u_0/12\beta)}(x - u_0/3t)].$$

From (4.30) it follows that the integral invariants $\int\langle u\rangle\,\mathrm{d}x$, $\int\langle u\rangle^2\,\mathrm{d}x$ do not change, while the form of the mean field does. At times $t \gg \beta/u_0 D$, the mean field acquires the form of a Gaussian pulse:

$$\langle u(x,t)\rangle = \frac{M_1}{\sqrt{(2\pi Dt)}}\exp\left\{-\frac{(x-u_0t/3)}{4Dt}\right\},$$

$$M_0 = \sqrt{12\beta u_0}\int_{-\infty}^{\infty}\mathrm{sech}^2\xi\,\mathrm{d}\xi. \tag{4.35}$$

One can see a dependence of the exact mean field on the initial amplitude. In the mean field *method*, however, memory of the initial amplitude completely vanishes.

Analysis of the asymptotic behavior shows that the mean field, for localized initial conditions of the stochastic KdV equation, disintegrates into a sequence of Gaussian pulses whose width increases as $\sqrt{t}$. Later on, in Refs. [4.10], [4.11], a number of analogous results for the stochastic KdV equation were obtained. The authors of those works considered the KdV external noise,

$$u_t - 6uu_x + u_{xxx} = \eta(t).$$

Applying the transformation

$$v(x,t) = u(x,t) - W(t), \qquad W = \int_0^t \eta(t')\,\mathrm{d}t',$$

This is turned into the KdV equation with stochastic parametric perturbation, in the form (4.34). Wadati used a cumbersome IST procedure as compared with the simple transformation proposed in Ref. [4.4].

The next model example is a stochastic NLS equation. Consider, first of all, a stochastic NLS equation of the following form [4.7]:

$$\begin{gathered} \mathrm{i}u_t + \tfrac{1}{2}u_{xx} + |u|^2 u = \varepsilon(t)u, \\ \langle\varepsilon\rangle = 0, \ \langle\varepsilon(t+\tau)\varepsilon(t)\rangle = 2\sigma_\varepsilon^2\delta(\tau). \end{gathered} \tag{4.36}$$

With the aid of the transformation

$$u = v\exp\left\{-\mathrm{i}\int_0^t \mathrm{d}t'\varepsilon(t')\right\}$$

this turns into a standard unperturbed NLS equation,

$$\mathrm{i}v_t + \tfrac{1}{2}v_{xx} + |v|^2 v = 0, \qquad v(t=0) = u_0(x). \tag{4.37}$$

and thus the exact mean is

$$|\langle u\rangle|^2 = 4\eta^2\,\mathrm{sech}^2 2\eta(x-\zeta)\exp(-\sigma_\varepsilon^2 t/2). \tag{4.38}$$

On the other hand, the application of the mean field method leads to

$$\langle u\rangle = u_{s0}(x,t)\exp(-\sigma_\varepsilon^2 t/4). \tag{4.39}$$

This is correct only at times $t \ll 1/\sigma_\varepsilon^2$. At large times it gives overestimated values, as compared with the exact solution.

A similar set of problems for perturbations of the form $R = \varepsilon(t)xq$ was examined in Refs. [4.7], [4.12]. It was also shown there that the solution is a decaying gaussian wave packet.

The third example is the problem of the transmission of an ultra-short light pulse in a resonantly absorbing medium when the density of resonant atoms varies along the propagation path in a random manner [4.13, 4.14]. In the case of exact resonance, and in the absence of inhomogeneous gain, the equaton has the form of a stochastic SG equation:

$$u_{xt}(x,t) = (-1 + r(t))\sin u,$$
$$\langle r(t_1)r(t_2)\rangle = 2\sigma^2\delta(t_1 - t_2). \tag{4.40}$$

Transformation of the variables reduces this equation to a standard SG equation. The exact mean field obtained in this way is (for $q = u_x$)

$$\langle q(x,t)\rangle = 2\eta\langle \operatorname{sech}[2\omega(t) + \zeta]\rangle,$$
$$\omega(t) = \int_0^t r(t')\,\mathrm{d}t', \qquad \zeta = 2\eta(x - x_0 - t/4\eta^2). \tag{4.41}$$

For $\sigma t \ll 4\eta^2$ we have

$$\langle q(x,t)\rangle = 2\eta \operatorname{sech}\zeta[1 - (\sigma t/4\eta^2)(1 - 2\tanh^2\zeta)].$$

A more precise result for $\langle q\rangle$ is

$$\langle q(x,t)\rangle = \left\{1 + \sum_{l=1}^{\infty}(-1)^l E_{21}(\pi/2)^l \frac{\mathrm{d}^l}{\mathrm{d}s^l}\right\}$$
$$\times \frac{\eta\sqrt{\pi}}{\sqrt{s}}\exp(-\zeta^2/4s) \approx \eta\sqrt{(\pi/\sigma t)}\exp(-\zeta^2/4\sigma t).$$

This is a consequence of Brownian motion of the soliton maximum. It can be also shown that the averaged soliton envelope is described by a linear diffusion equation.

4.3 ADIABATIC DESCRIPTION OF SOLITON DYNAMICS IN RANDOM FIELDS

Let us consider a statistical adiabatic approximation appearing in the analysis of stochastic nonlinear wave equations of the form

$$\hat{N}Lq = R[\varepsilon(x,t), q, q_t, q_x, \ldots]. \tag{4.42}$$

Here $\hat{N}L$ is a nonlinear operator. For the KdV it equals

$$\hat{N}L \equiv \partial_t + 6q\partial_x + \partial_{xxx},$$

for the NLS it is

$$i\partial_x + \partial_{xx} + 2|q|^2,$$

and for the SG equation it is

$$\partial_{tt} - \partial_{xx} + \sin(\cdots).$$

R is a perturbation operator, $\varepsilon(x, t)$ is a random function to be assumed later to be Gaussian with zero mean

$$\langle \varepsilon \rangle = 0, \quad \langle \varepsilon(x, t)\varepsilon(y, \tau) \rangle = B_{l_\varepsilon, \tau_\varepsilon}(x - y, t - \tau). \tag{4.42a}$$

Here l_ε is a correlation length, τ_ε a correlation time. For $l_\varepsilon, \tau_\varepsilon \to 0$,

$$B(x - y, t - \tau) \to 2\sigma_e^2 \delta(x - y)\delta(t - \tau),$$

that corresponds to the model of a δ-correlated random process.

Let $\varepsilon(x, t)$ be a weak random perturbation, i.e. $\varepsilon \ll 1$. It is reasonable to assume that the soliton parameters (amplitude, velocity, phase, etc.) vary smoothly in time while their mutual relationships remain the same as in a free soliton. Below we will consider some typical cases; stochastic dynamics of NLS, KdV, and SG solitons under random perturbations.

4.3.1 Dynamics of NLS Solitons in the Field of Fluctuations

Consider the problem of the evolution of a one-soliton initial state under the action of a random field $\varepsilon(t)$ and damping Γ [4.15], [4.16],

$$iq_t + \tfrac{1}{2}q_{xx} + |q|^2 q = -i\Gamma q + \varepsilon(t). \tag{4.43}$$

with the aid of perturbation theory (see Appendix 1), we obtain in the adiabatic approximation a set of equations for soliton parameters,

$$\frac{d\eta}{dt} = -2\Gamma\eta - \frac{\pi\varepsilon(t)\sin\delta(t)}{2\cosh[\pi\mu(t)/2\eta(t)]}, \tag{4.44}$$

$$\frac{d\xi}{dt} = 2\mu - \frac{\pi^2\varepsilon(t)\cos\delta(t)\sinh[\pi\mu(t)/2\eta(t)]}{8\eta^2\cosh^2[\pi\mu/2\eta]}, \tag{4.45}$$

$$\frac{d\mu}{dt} = \frac{\pi\varepsilon\mu \sin\delta(t)}{4\eta \cosh[\pi\mu/2\eta]}, \tag{4.46}$$

$$\frac{d\delta}{dt} = 2(\mu^2 + \eta^2) - \frac{\pi\varepsilon \cosh\delta(t)}{\pi \cosh[\pi\mu/2\eta]}. \tag{4.47}$$

Now we will represent the parameters in the forms

$$\mu = \mu_0 + \mu_1 + \cdots, \quad \delta = \delta_0 + \delta_1 + \cdots, \quad \eta = \eta_0 + \eta_1 + \cdots,$$
$$\xi = \xi_0 + \xi_1 + \cdots; \quad \mu_1, \eta_1, \delta_1, \xi_1 \sim \varepsilon.$$

Here

$$\mu_0 = \text{constant}, \quad \eta_0 = \eta_i \exp(-2\Gamma t), \quad \eta_i = \eta_0(t=0),$$
$$\delta_0 = 2\mu_0^2 t + \delta_i + \eta_i^2(\exp(-2\Gamma t) - 1)/\Gamma, \quad \xi_0 = 2\mu_0 t + \xi_0.$$

Let us write out these formulas for the case $\Gamma = 0$:

$$\frac{d\eta}{dt} = -\frac{\pi\varepsilon(t)\sin\delta_0(t)}{2\cosh[\pi\mu_0/2\eta_0]}.$$

from this, for the r.m.s. soliton amplitude fluctuation we find (for $\Gamma = 0$) that

$$\langle \eta_1^2 \rangle = \frac{\pi\sigma_\varepsilon^2 t}{\cosh[\pi\mu_0/2\eta_0]} \left\{ 1 - \frac{\sin[4(\mu_0^2 + \eta_0^2)t]}{4(\mu_0^2 + \eta_0^2)t} \right\}. \tag{4.48}$$

Taking account of $\tau_\varepsilon \neq 0$ leads to small-parameter correction $\tau_\varepsilon/t_s \ll 1$ corrections. As is seen from the last expression, stochastic growth of soliton energy is caused by the interaction with the fluctuating field, the linear amplitude increase with respect to t being accompanied by oscillations of frequency $4(\mu_0^2 + \eta_0^2)$. For times $4(\mu_0^2 + \eta_0^2)t \gg 1$, we obtain for the diffusion coefficient

$$D_\eta = \lim_{t \to \infty} \frac{\langle \eta_1^2 \rangle}{t} = \frac{\pi\sigma_\varepsilon^2}{4\cosh^2(\pi\mu_0/2\eta_0)}.$$

In a similar way we find the mean values of $\langle \mu_1^2 \rangle, \langle \xi_1^2 \rangle, \langle \delta_1^2 \rangle$.

4.3.2 Stochastic Parametric Resonance of NLS Solitons

Another model example is that of parametric interaction between NLS solitons and a random field. The initial equation, describing NLS soliton propagation in a medium with fluctuating parameters, has the following form [4.17]:

$$\mathrm{i}q_t + \tfrac{1}{2}q_{xx} + |q|^2 q = -\omega^2(t)x^2 q. \tag{4.49}$$

It describes, for example, the propagation of a Langmuir soliton in plasma with fluctuating density. For $\omega = \omega_0$ the soliton describes periodic oscillations around the point $x = 0$. In the adiabatic approximation we obtain equations for the soliton parameters,

$$\mathrm{d}\xi/\mathrm{d}t = 2\mu, \quad \mathrm{d}\eta/\mathrm{d}t = 0, \quad \mathrm{d}\mu/\mathrm{d}t = \omega^2(t)\xi/2. \tag{4.50}$$

For the soliton center coordinate we have now

$$\mathrm{d}^2\xi/\mathrm{d}t^2 + \omega^2(t)\xi = 0. \tag{4.51}$$

Let $\omega^2(t) = \omega_0^2(1 + \varepsilon(t))$, where $\varepsilon(t)$ is a random function satisfying (4.42a). Relying on the results of Ref. [4.8], the mean values of $\langle\xi\rangle, \langle y\rangle (y = \dot{\xi})$ can be shown to satisfy the set of equations

$$\mathrm{d}\langle\xi\rangle/\mathrm{d}t = \langle y\rangle, \mathrm{d}\langle y\rangle/\mathrm{d}t = -\omega_0^2\langle\xi\rangle.$$

Under the initial conditions $\xi(0) = 0, y(0) = 1$, the solution of this set will be

$$\langle\xi\rangle = (\sin\omega_0 t)/\omega_0, \quad \langle y\rangle = \cos\omega_0 t.$$

The second moments are described by the equations

$$\langle\dot{\xi}^2\rangle = 2\langle y\rangle, \quad \langle\dot{\xi y}\rangle = \langle y\rangle^2 - \omega_0^2\langle\xi^2\rangle,$$
$$\langle\dot{y}^2\rangle = -2\omega_0^2\langle\xi y\rangle + 2\sigma_\varepsilon^2\omega_0^4\langle\dot{\xi}^2\rangle,$$

with the conditions

$$\langle\xi^2\rangle = \langle\xi y\rangle = 0, \qquad \langle y^2\rangle = 1, \qquad t = 0.$$

The solution of this set of equations for $\sigma^2\omega_0$ is

$$\langle\xi^2(t)\rangle = (2\omega_0^2)^{-1}[\exp(\sigma^2\omega_0^2 t) - \exp(-\sigma^2\omega_0^2 t/2)(\cos 2\omega_0 t$$
$$+ \tfrac{3}{4}\sigma^2\omega_0\sin\omega_0 t)].$$

For $t \sim (\sigma_\varepsilon^2)^{-1}$ the soliton is seen to accelerate stochastically due to fluctuations of the system parameters. The time of acceleration is

$$t \gg t_s, t_s \sim 1/2\eta v_s, t = 1/\sigma_\varepsilon^2\omega_0^2.$$

Thus, while solving this problem we encountered a new phenomenon—*stochastic parametric soliton resonance.* As shown by the analysis, this is of universal character and exists in other nonlinear wave systems.

Let us consider another example to illustrate this statement,

$$iq_x + \tfrac{1}{2}q_{\tau\tau} + |q|^2 q + i\gamma q - i\varepsilon(x,\tau)\exp(-2\gamma_1 x)q = 0. \qquad (4.52)$$

This equation occurs in the problem of soliton propagation in a single-mode fiber with fluctuating pumping [4.17, 4.18]. Here q is a dimensionless amplitude, $\tau = t - x/v_g$ is the retarded time, x is the direction of pulse propagation, γ_1, γ are increments due to damping on the Stokes and pump frequencies, respectively, ε is a random function defined by the field of partially coherent pumping and medium properties. In the following, for the sake of simplifying the final formulas, we will assume that ε is a function of the coordinate x only.

Equation (4.52) turns, for $\gamma = \varepsilon = 0$, into the NLS with one-soliton solution

$$q_s = 2\eta \operatorname{sech} z \exp[i\mu z/\eta + i\delta], \qquad z = 2\eta(\tau - \xi).$$

The random function is represented in the form $\varepsilon(x) = \varepsilon_0 + \tilde{\varepsilon}(x)$, where $\tilde{\varepsilon}(x)$ is a Gaussian random function. Applying the adiabatic approximation of perturbation theory for solitons, we obtain for the soliton amplitude

$$d\eta/dx = 2\eta[\varepsilon_0 \exp(-2\gamma_1 x) - \gamma + \tilde{\varepsilon}\exp(-2\gamma_1 x)]. \qquad (4.53)$$

Equation (4.53) is solved for the initial condition $\eta = \eta_0$ to give the form

$$\begin{aligned} &\eta = \eta_0 \exp(I_0 + I_1); \\ &I_0 = (\varepsilon/\gamma_1)(1 - \exp(-2\gamma_1 x)) - 2\gamma_1 x; \\ &I_1 = 2\int_{-\infty}^{\infty} \varepsilon(x)\exp(-2\gamma_1 x)dx. \end{aligned} \qquad (4.54)$$

Averaging of (4.54) over all realized random values of the mean soliton amplitude yields

$$\langle \eta(x) \rangle = \eta_0 \exp[I_0 + \tfrac{1}{2}\langle I_1^2 \rangle]. \qquad (4.55)$$

For instance, in the case of a δ-correlated random process ε, we have

$$\eta(x) = \eta_0 \exp[2(2\sigma_\varepsilon^2 + \varepsilon_0 -)x - 2\gamma_1(\varepsilon_0 + 4\sigma_\varepsilon^2)x^2]. \qquad (4.56)$$

The analysis of (4.55) indicates the existence of a range of parameters, in which $\langle \eta(x) \rangle$ grows with its propagation along fiber because of the energy pumping into the soliton out of the random field, i.e. a phenomenon of stochastic parametric resonance occurs. It is especially clearly seen for $\varepsilon_0 = \gamma$, when dissipation loss is precisely compensated by the interaction with the regular part of the pumping wave. In this case, ignoring the dissipation of the pumping wave as well, $\gamma_1 = 0$, we derive from (4.55) that $\langle \eta \rangle = \eta_0 \exp(4\sigma_\varepsilon^2 x)$. As in evident, the growth of the mean amplitude is due to the fluctuations of ε. This is in agreement with numerical calculations carried out in Ref. [4.18], where the interaction between an SRS soliton and a noise pumping field was studied.

Let us also give the formula for the probability density function of solitons over amplitudes. It is

$$P_x(\eta) = \frac{1}{\sqrt{(2\pi)\langle I_1^2 \rangle}\,\eta} \exp\left(- \frac{[\ln(\eta/\eta_0) - I_0]^2}{2\langle I_1^2 \rangle} \right\}; \tag{4.57}$$

$$\langle I_1 \rangle = 0, \langle I_1^2 \rangle = 2\sigma_\varepsilon^2[1 - \exp(-4\gamma_1 x)]/\gamma_1 = 8\sigma_\varepsilon^2 x(1 - 2\gamma_1 x). \tag{4.58}$$

The soliton amplitudes are seen to follow a log-normal distribution.

4.3.3 Stochastic Parametric Resonance of KdV Solitons

Let us consider the evolution of KdV solitons under nonstationary fluctuations of the medium parameters, and under dissipation. The wave equation has the form [4.17]

$$u_t - 6uu_x + u_{xxx} = \varepsilon(t)u - \gamma u_{xx}, \tag{4.59}$$

where $\varepsilon(t)$ satisfies the condition (4.42a). The one-soliton solution is chosen in the form

$$u_s(x, t) = -2\kappa^2 \operatorname{sech}^2[\kappa(x - 4\kappa^2 t)].$$

It is readily seen that for equation (4.59) the following integral relation is valid:

$$\frac{d}{dt}\int_{-\infty}^{\infty} u^2 dx = \varepsilon(t)\int_{-\infty}^{\infty} u^2 dx + \gamma \int_{-\infty}^{\infty} (u_x)^2 \, dx. \tag{4.60}$$

The soliton energy is

$$E_s = \int_{-\infty}^{\infty} u^2 dx = \tfrac{16}{3}\kappa^3.$$

From (4.60) we derive

$$\frac{d\kappa}{dt} = \tfrac{2}{3}\varepsilon(t)\kappa - \tfrac{8}{15}\gamma\kappa^3, \tag{4.61}$$

whose solution is

$$\kappa(t) = \kappa_0 \frac{\exp(I(t))}{[1 + \frac{16}{15}\kappa_0^2\gamma \int_0^t dt' \exp(2I(t'))]^{1/2}},$$

$$I(t) = \tfrac{2}{3}\int_0^t \varepsilon(\tau)d\tau.$$

Let us analyze the behavior of the mean energy for different values of the parameter $\varepsilon(t) = \varepsilon_0 + \tilde{\varepsilon}$. For $\varepsilon_0 = 0$ the soliton energy increases due to the interaction with the random field, i.e. stochastic parametric resonance is observed:

$$\langle \kappa^2(t) \rangle = \kappa_0^2 \exp(\tfrac{16}{9}\sigma^2 t).$$

Taking the next term of the expansion into account shows that its contribution reduces the energy increase. The resulting expression has the form

$$\langle \kappa^2 \rangle = \kappa_0^2 \exp(\tfrac{16}{9}\sigma^2 t)\{1 - \kappa_0^2\gamma/5\sigma^2[\exp(\tfrac{16}{3}\sigma^2 t) - 1]\}.$$

It is seen that for $\sigma^2 = \frac{3}{5}\kappa_0^2\gamma$ a stationary behavior is possible.

Consider then the problem of the soliton amplitude density function. By a standard procedure, we obtain the Fokker–Planck equation for $P_t(\kappa)$

$$\frac{\partial P_t}{\partial t} = -\frac{\partial}{\partial \kappa}[f(\kappa)P_t] + \sigma^2 \frac{\partial}{\partial \kappa}\left\{g(\kappa)\frac{\partial}{\partial \kappa}[g(\kappa)P_t]\right\};$$

$$f(\kappa) = \tfrac{2}{3}\varepsilon_0\kappa - \tfrac{8}{15}\gamma\kappa^3, \qquad g(\kappa) = \tfrac{2}{3}\kappa.$$

This has the stationary solution

$$P(\kappa) = 2\beta^\alpha \exp(-\beta\kappa^2)\kappa^{2\alpha-1}, \qquad \alpha = 3\varepsilon/4\sigma^2, \quad \beta = 3\gamma/5\sigma^2,$$

and we can see, in the range of parameters $2\alpha > 1$, that the density function has its maximum for the amplitude

$$\kappa = [(2\alpha - 1)/2\beta]^{1/2}.$$

Note that the effect of the KdV equation, for normally distributed noise sources and for dissipation, was studied by Meerson [4.19], who found the presence of a plateau in the soliton energy density function $P(E, t)$. The possibility of stochastic heating of the KdV soliton was first discussed in Ref. [4.20].

4.3.4 Stochastic Parametric Dynamics of the SB Solitons

We consider now the motion of the SG soliton in a system with fluctuating parameters [4.21],

$$u_{tt} - u_{xx} + \sin u = -\varepsilon f(x,t) u_{xx} - \Gamma u_t, \tag{4.62}$$

where $f(x,t)$ is a Gaussian Random process. From equation (4.62) we find the set of equations for the soliton coordinate $\xi(t)$ and velocity $v(t)$ as

$$\mathrm{d}v/\mathrm{d}t = -\Gamma v - \varepsilon/4 \int_{-\infty}^{\infty} f(x,t) u_{xx}(z) \operatorname{sech} z \, \mathrm{d}z, \tag{4.63}$$

$$\mathrm{d}\xi/\mathrm{d}t = v - (\varepsilon/4) \int_{-\infty}^{\infty} f(x,t) u_{xx} z \operatorname{sech} z \, \mathrm{d}z; \tag{4.64}$$

$$z = (x - \zeta(t))/\sqrt{(1 - v^2)}.$$

These are integrodifferential stochastic equations.

Now let us determine the probability density for the soliton coordinate and velocity, following the technique of Markov Chain approximations [4.8]. The set of equations is of the first order with respect to t, and the initial conditions are defined for $t = 0$. Hence, the functions v and ξ are functionally dependent only on the previous values with respect to t. Then a causality condition exists,

$$f_1(v,t) = \varepsilon/4 \int_{-\infty}^{\infty} f(x,t) u_{xx}(z) \operatorname{sech} z \, \mathrm{d}z; \tag{4.65}$$

$$\langle f_1(v,t) f_1(v',t') \rangle = 2\delta(t - t') D(v, v'),$$

$$\delta v(t)/\delta f_1(t') = 0 \qquad \text{for} \quad t < 0, \quad t > t'.$$

To derive an equation for the probability density $P(v, \xi; t)$ we need to know the derivative $\delta v(t)/\delta f_1(v, t')$ for $t < t'$, $t = t'$. For this, (4.63) is integrated over t and the resulting expression is acted upon by the variational derivative $\delta/\delta f$ to yield

$$\frac{\delta v(t)}{\delta f_1(v,t')} = \delta(v - v(t')) - \int_{t'}^{t} \mathrm{d}\tau \int_{-\infty}^{\infty} \mathrm{d}v' [-\Gamma v' - f(v',\tau)]\delta(v' - v(\tau)) \frac{\delta v(\tau)}{\delta f_1(v',\tau)}. \tag{4.66}$$

Here the causality condition (4.65) has been taken into account. For $t \to t'$ we have

$$\lim_{t \to t'} \frac{\delta v(t)}{\delta f_1(t')} = \delta(v - v(t)). \tag{4.67}$$

Now we introduce the joint probability density

$$P(v, \zeta; t) = \langle \delta(v - v(t)) \rangle \langle \delta(\zeta - \zeta_0(t) \rangle, \tag{4.68}$$

where $v(t), \zeta_0(t)$ is a solution that refers to a specified realization of $f(v, t)$, averaging being done over the set of all realizations of $f(v, t)$. Equation (4.68) is differentiated to yield

$$\frac{\partial P_t}{\partial t} = -\frac{\partial}{\partial v}(\Gamma v P_t) - v\frac{\partial P_t}{\partial \zeta} - \frac{\partial}{\partial v} \langle f_1(v,t)\delta(v - v(t)) \rangle. \tag{4.69}$$

in order to evaluate the correlator $\langle f_1 \delta(v - v(t)) \rangle$, the Furutsu–Novikov equation is used,

$$\langle f_1 \delta(v - v(t)) \rangle = \int_0^t \mathrm{d}t' \int_{-\infty}^{\infty} \mathrm{d}v' \langle f_1(v,t) f_1(v',t) \rangle \left\langle \frac{\delta}{\delta f_1(t')} \delta(v' - v) \right\rangle. \tag{4.70}$$

Then the causality condition (4.67) may be applied to (4.70) to lead eventually to the Fokker–Planck equation:

$$\frac{\partial P_t}{\partial t} = -\frac{\partial P_t}{\partial \zeta_0} + \frac{\partial}{\partial v}(\Gamma v P_t) + \frac{\partial}{\partial v}\left(D(v) \frac{\partial P_t}{\partial v} \right). \tag{4.71}$$

Its initial condition is chosen in the form

$$P(\zeta_0, v; t)_{t=0} = \delta(v - v_0)\delta(\zeta_0),$$

$D(v)$ being the diffusion coefficient, equal to

$$D(v) = \frac{(2\pi)^3 \sigma_\varepsilon^2}{32} \int_0^{\infty} k^2 \varphi^2(k) S(k, kv)\, \mathrm{d}k$$

$$= \frac{(2\pi)^3\sigma_\varepsilon^2}{32}\int_0^\infty \frac{k^4 \mathrm{d}k}{(1+k^2v^2\tau_\varepsilon^2)(1+k^2l_\varepsilon^2)\sinh^2(\pi k/2)}. \tag{4.72}$$

Let us examine the case $\tau_\varepsilon = l_\varepsilon = 0$.

A stationary velocity distribution has the form

$$P_{\mathrm{st}} = \sqrt{(2\Gamma/\pi D)}\exp(-\Gamma v^2/D), \qquad D = 96\sigma_\varepsilon^2. \tag{4.73}$$

We will also supply the results of calculation using a nonstationary density function for the case of large-scale fluctuations [4.22], i.e. for $l \gg l_{\mathrm{s}} = (1 - v_0^2)^{1/2}$, where l_{s} is the width of the unperturbed soliton (kink). In this case a soliton in the time-fluctuating field manages to adapt itself to spatial nonuniformities of the system. The mean velocity and coordinate of soliton are evaluated as follows:

$$\langle v \rangle = v_0 - 3\pi^{3/2}2^{-1/2}v_0(1 - v_0^2)^2 t(\nu l)^{-1},$$
$$\langle x \rangle = v_0 t - 3\pi^{3/2}2^{-1/2}v_0(1 - v_0^2)^2 t^2(\nu l)^{-1}, \qquad \nu = 16\varepsilon^{-2}.$$

We can see that the soliton in a medium having large-scale inhomogeneities slows down, most noticeably in the randomly inhomogeneous region. Note that the adiabatic approximation for the KdV equation, for instance, does not fully describe soliton dynamics, since under the perturbation a continuous spectrum is excited (a 'tail' grows), so the results ought to be corrected for this effect. In the case of SG-soliton motion in a random potential, linear waves that can change the adiabatic dynamics (the so-called 'radiation reaction' effect) are also excited. However, the contribution of the radiation to the dynamics will be of order ε^3 and can be neglected. This problem will be studied in Section 4.4.

4.4 STATISTICAL BORN APPROXIMATION

The method expounded in this section is based on the expansion of the initial solution in series close to the solution of the unperturbed problem. Here we will isolate two methods: the Green function method and IST-based perturbation theory.

Let us examine the application of the first method to the example of the domain wall motion in randomly inhomogeneous quasi-one-dimensional ferroelectrics [4.23]. The equation of motion for the shift fields $u(x, t)$ is

$$mu_{tt} - mc_0^2 u_{xx} - Au + Bu^3 = -V(x)u. \tag{4.74}$$

For $V(x)=0$ the equation of motion turns into that considered by Aubry [4.24], and Krumhansl and Schrieffer [4.25]. The solution for a domain wall has the form

$$u_s(x,t)=u_0\tanh[\gamma(x-vt)/(\sqrt{2})\xi_0]. \tag{4.75}$$

where

$$u_0^2=A/2B,\quad \xi_0^2=mc_0^2/A,\quad \gamma^{-1}=\sqrt{(1-\beta^2)},\quad \beta=v/c_0,$$

and where v is the velocity of the domain wall.

Now we proceed with dimensionless variables in equation (4.74), setting

$$\tau=\omega_0 t, z=x/\xi_0, \varepsilon(z)=V/A, \Phi=u/u_0, \omega_0^2=A/m.$$

Instead of (4.74) we then obtain the following equation:

$$\Phi_{\tau\tau}-\Phi_{zz}-\Phi+\Phi^3=-\varepsilon(z)\Phi. \tag{4.76}$$

Let us assume that the perturbation is weak, and seek a solution of (4.76) in the form

$$\Phi=\Phi_s+\Phi_1+\Phi_2+\cdots, \Phi_1\sim\varepsilon, \Phi_2\sim\varepsilon^2,\ldots.$$

The linearized equation for the first-order correction is written in the form

$$\Phi_{1\tau\tau}-\Phi_{1zz}+2[1-\tfrac{3}{2}\operatorname{sech}^2\gamma\sqrt{2}(z-\beta\tau)]\Phi_1=\varepsilon(z)\tanh[\gamma(z-\beta\tau)/\sqrt{2}]. \tag{4.77}$$

In the following it is convenient to proceed to a reference frame moving together with the domain wall. Thus set

$$z'=\gamma(z-\beta\tau),\qquad \tau'=\gamma(\tau-\beta\tau).$$

As a result the following equation for Φ_1 is obtained:

$$\Phi_{1\tau\tau}-\Phi_{1z'z'}+2(1-\tfrac{3}{2}\operatorname{sech}^2 z'/\sqrt{2})\Phi_1=-\varepsilon[\gamma(z'+\beta\tau)]\tanh(z'/\sqrt{2}). \tag{4.78}$$

Then the Fourier transformation of (4.78) over τ' yields

$$\Phi_{1z'z'}(z',\omega)+[\omega^2-2(1-\tfrac{3}{2}\operatorname{sech}^2 z'/\sqrt{2})]\Phi_1(z',\omega)=F(z',\omega), \tag{4.79}$$

where

$$F=\frac{1}{\gamma\beta}\tilde{\varepsilon}(\omega/\gamma\beta)\tanh(z'/\sqrt{2})\exp(\mathrm{i}\omega z'/\beta)$$

and $\tilde{\varepsilon}$ is the Fourier transform of the random function, and satisfies the following relations:

$$\langle \tilde{\varepsilon} \rangle = 0, \quad \langle \tilde{\varepsilon}(k)\tilde{\varepsilon}(k') \rangle = 2\pi\sigma_v^2 \delta(k+k')/\xi_0 A^2. \tag{4.80}$$

Equation (4.79) for $\varepsilon = 0$ is the Schrödinger equation with a reflectionless potential, whose spectrum and eigenfunctions are well known [4.26], namely

$$\begin{aligned}
&\omega^2 = 0, \quad \psi_0(z') = \sqrt{(3/4\sqrt{2})}(\operatorname{sech}^2 z')/\sqrt{2},\\
&\omega^2 = \tfrac{3}{2}, \quad \psi_1(z') = \sqrt{(3/2\sqrt{2})}\operatorname{sech}^2 z'/\sqrt{2}\sinh^2 z'/\sqrt{2},\\
&\omega^2 = 2 + q^2 = \omega_q^2,\\
&\psi_q = [\sqrt{(2\pi)}(2 - 2q^2 - (\mathrm{i}\sqrt{2})3q)]^{-1}\\
&\qquad \times (3\tanh^2(z'/\sqrt{2}) - 3\mathrm{i}\sqrt{2}q\tanh(z'/\sqrt{2}) - 1 - 2q^2).
\end{aligned} \tag{4.81}$$

Now equation (4.79) will be solved with the Green function formed from the eigenfunctions of (4.81), to lead to

$$\Phi_1(z',\omega) = \int_{-\infty}^{\infty} G_\omega(z',z'_1)F(z'_1,\omega)\,\mathrm{d}z'_1,$$

where

$$G_\omega(z',z'_1) = \frac{\mathrm{i}}{2k}\begin{cases}\psi_k(z')\psi_k^*(z'_1), & z' > z'_1\\ \psi_k^*(z')\psi_k(z'_1), & z' < z'_1.\end{cases}$$

$$\begin{aligned}
k &= \sqrt{(2-\omega^2)} \quad \text{for} \quad |\omega| \leqslant \sqrt{2},\\
k &= -\mathrm{i}\sqrt{(\omega^2 - 2)} \quad \text{for} \quad \omega > \sqrt{2},\\
k &= \mathrm{i}\sqrt{(\omega^2 - 2)} \quad \text{for} \quad \omega > -\sqrt{2}.
\end{aligned}$$

I-n the following, we need to distinguish between the linear mode contributions, namely:

$$\Phi_1(z',\omega) = \Phi_{10}(z',\omega) + \Phi_{11}(z',\omega) + \Phi_{1c}(z',\omega).$$

It is readily seen that

$$\begin{aligned}
\Phi_{10} &= \mathrm{i}(3\pi/4\beta^3\gamma)\varepsilon(\omega/\beta\gamma)\operatorname{cosech}(\pi\omega/\sqrt{2}\beta)\operatorname{sech}^2(z'/\sqrt{2}),\\
\Phi_{11} &= (3\pi/2\beta^3\gamma)\varepsilon(\omega/\beta\gamma)[(2\omega^2 - \beta^2)/(2\omega^2 - 3)]\\
&\qquad \times \operatorname{sech}(\pi\omega\beta\sqrt{2})\operatorname{sech}(z'/\sqrt{2})\sinh(z'/\sqrt{2}),\\
\Phi_{1c} &= (3\sqrt{2}\pi/\beta^2\gamma)\varepsilon(\omega/\beta\gamma)\omega^2(1 - \beta^2/3)[2k(2 - 2k^2 + 3\mathrm{i}\sqrt{2}k]^{-1}\\
&\qquad \times \operatorname{cosech}(\pi(\omega/\beta - k)/\sqrt{2})\exp(\mathrm{i}kz'),
\end{aligned}$$

where Φ_{10}, Φ_{11} correspond to the discrete spectrum and Φ_{1c} to the continuous one, i.e. linear wave excitation. For experimental purposes, a correlation function $S(q, \omega)$, describing neutron or light scattering, is useful. The bar means averaging over configurations ε. In order to evaluate this correlation function, we will make use of the phenomenological statistical mechanics of ideal gas for the domain walls. We will confine ourselves to the study of a collisionless mode, i.e. consider a free motion of domain walls corresponding to the case which is linear in soliton density ($n_s \ll 1$). The statistical mechanics of domain walls applied in this section is valid at temperatures $T > k_B^{-1}$. For our model there exists another characteristic temperature which imposes a lower limit. Indeed, at very low temperature, bound states of domain walls appear that can make the main contribution to the statistics.

A measure for the scattering intensity is the structure function. Insertion of the expansion and subsequent averaging yields the following expression for $S(q, \omega)$:

$$S_{\mathrm{do}}(q, \omega) = \frac{(3\sqrt{3})\pi\sigma\theta}{\pi\omega\theta}\left[\frac{\pi}{\sqrt{2}}(q + \omega(\sqrt{2\pi\omega\theta}))^{1/3}\right]^2$$
$$\times \operatorname{cosech}^2\left[\frac{\pi}{\sqrt{2}}(q + \omega(\sqrt{2\pi\omega\theta}))^{1/3}\right]$$
$$\times \exp\left[-2\pi q - \tfrac{3}{2}\left(2\pi^2\frac{\omega^2}{\theta}\right)^{2/3}\right],$$

where $\theta = k_B T/mc_0^2$.

One can see that the term $S_{\mathrm{do}}(q, \omega)$ describes the effect of impurities on the central peak intensity. The temperature behavior of S_{do} is of δ-function character for frequency $\omega \to 0$, i.e. for $T \to 0$, whereas $S_0(q, \omega)$ decreases to zero. Growth of the central peak intensity with T seems to be related to the fact that the dynamics of the domain wall gas in the field of a random potential is equivalent to a diffusive motion of the domain wall, i.e. at a reduced temperature we can see the progression from a collisionless mode to a diffusive one. Then at the reduced temperature, the height of the central peak increases, while its width decreases.

4.4.1 IST-based Statistical Perturbation Theory

As mentioned above, it is natural to apply an IST-based perturbation theory for analyzing nearly integrable systems, including statistical perturbations.

Let us examine the propagation of the SG soliton in a fluctuating-parameter medium [4.27]. The wave equation describing the motion of the SG soliton in the random potential field is chosen in the form

$$u_{tt} - u_{xx} + \sin(u) = \varepsilon(x)R(u), \tag{4.82}$$

where $\varepsilon(x)$ describes a random medium inhomogeneity. A soliton solution of (4.82) for $\varepsilon = 0$ has the form

$$u_1(x,t) = 4\tan^{-1}\exp\left(\frac{x-\zeta}{\sqrt{(1-v^2)}}\right), \qquad \zeta = \zeta_0 + vt.$$

Previously it was shown that a soliton moving in the random potential field behaves, in the adiabatic approximation, as a classical particle. Here we will examine another aspect of the problem associated with the soliton-radiated waves in the randomly inhomogeneous medium.

According to the IST, the radiation energy can be calculated with the aid of the formula (see Appendix 2)

$$E = \frac{4}{\pi}\int_{-\infty}^{\infty} \mathrm{d}\lambda \left(1 + \frac{1}{4\lambda^2}\right)|b(\lambda,t)|^2, \tag{4.83}$$

where the coefficient $b(\lambda, t)$ can be derived from the equation

$$\frac{\mathrm{d}b}{\mathrm{d}t} = 2\mathrm{i}\omega(\lambda)B + \frac{i}{4}\int_{-\infty}^{\infty} R(u_s)\varepsilon(x)[\varphi_1(x,t)\psi_2^*(x,t) + \varphi_2(x,t)\psi_1^*(x,t)]. \tag{4.84}$$

Here $R(u_s)$ describes the perturbation, and ψ and φ are the Jost functions of the unperturbed sine–Gordon equation. From (4.83) one infers that

$$\frac{\mathrm{d}E}{\mathrm{d}t} = \frac{16}{\pi}\int_0^{\infty} \mathrm{d}\lambda \frac{\omega(\lambda)}{\lambda} \mathrm{Re}\left[b^* \frac{\mathrm{d}b}{\mathrm{d}t}\right], \tag{4.85}$$

where

$$\omega(\lambda) = \tfrac{1}{2}(\lambda + 1/4\lambda).$$

Solving (4.84) and making use of (4.85), one can obtain an expression for soliton radiation power (in the case $R(u) = u_{xx}$) as

$$\frac{\mathrm{d}\langle E\rangle}{\mathrm{d}t} = \frac{2\alpha^2}{\pi^3\gamma^2}\int_1^{\infty} \frac{\mathrm{d}\omega}{k}\frac{f(\omega)}{\cosh^2(\pi\gamma\omega/2v_1)}. \tag{4.86}$$

The case $R(u) = \sin u$ has been considered in Ref. [4.50] where

$$f(\omega) = \frac{\left[\left(v_1^2 - \frac{(\omega+k)^2}{4}\right)\frac{\gamma\omega}{v} - \frac{v_1}{2}(\omega+k)\left(1 - \frac{\omega^2\gamma^2}{v^2}\right)\right]^2}{(\lambda^2 + v_1^2)(1 + \chi_0^2 l_\varepsilon^2)},$$

$$\chi_0^2 = k + \frac{\omega}{v}, \qquad v_1 = \frac{1}{2}\sqrt{\left(\frac{1+v}{1-v}\right)},$$

$$k = \sqrt{(\omega^2 - 1)}, \qquad \langle \varepsilon(x)\varepsilon(x')\rangle = \frac{\alpha^2}{2l_\varepsilon}\exp(-|x-x'|/l_\varepsilon),$$

α is the power of the random potential, and

$$\gamma = \sqrt{(1 - v^2)}.$$

As seen from (4.86), the radiation spectrum comprises all the frequencies starting from the threshold one at $\omega = 1$. For $v \to 0$, the integral in (4.86) allows the estimate

$$\frac{d\langle E\rangle}{dt} = \frac{\alpha^2 \exp(-\pi/v)}{(\sqrt{2})v^{7/2}\pi^{5/2}(1 + l_\varepsilon^2/v^2)}. \tag{4.87}$$

This shows that the soliton radiation power in a randomly inhomogeneous medium is exponentially small for low velocities. For $v \to 1$ we obtain

$$\frac{d\langle E\rangle}{dt} \sim \frac{2\alpha^2}{\pi^3(1 - v^2)}, \tag{4.88}$$

i.e. the radiation power grows. It should be noted that according to Refs. [4.28], [4.29], part of the radiated energy can be consumed to generate small-amplitude breathers.

In a similar way we can consider the radiation of waves by solitons under random perturbations in other systems as well, that are also described by nearly integrable equations.

Now we will report some results for the KdV and NLS solitons in random fields, following Ref. [4.30].

Let us choose a perturbed KdV equation in the form

$$u_t - 6uu_x + u_{xxx} = \varepsilon f(x,t)R(u). \tag{4.89}$$

We will examine statistical features of the radiation field for some special types of perturbation. The correlation function for the random function has the form

$$\langle f(x,t)f(x',t')\rangle = D_T(t-t')B_L(x-x'), \qquad \langle f\rangle = 0.$$

Here L and T indicate a correlation radius and time. It is believed that the Fourier transform $d_T(\omega)$ is a real function. The analog of (4.83) for the mean power emitted per unit of time by the KdV soliton is of the form

$$P = \frac{\mathrm{d}}{\mathrm{d}t}\langle E_{\mathrm{rad}}\rangle = \int_0^\infty \langle p(k)\rangle \,\mathrm{d}k = \frac{32}{\pi}\int_0^\infty k^4 \operatorname{Re}\left\langle b(k,t)\frac{\partial b^*(k,t)}{\partial t}\right\rangle \mathrm{d}k. \qquad (4.90)$$

The Jost coefficient is defined from the equation (A.90)

$$\frac{\partial b(k,t)}{\partial t} = 8\mathrm{i}k^3 b(k,t) - \frac{\mathrm{i}\varepsilon \exp(-2\mathrm{i}\kappa\xi)}{2\kappa k(k^2+\kappa^2)} \times \int_\infty^\infty \mathrm{d}z\, R[u_s(z)](k - \mathrm{i}\kappa \tanh z)^2 \exp(-2\mathrm{i}kz/\kappa). \qquad (4.91)$$

From this one can derive $b(k,t)$ for $t \gg t_s \sim 1$, and further the correlator $\langle b(k,t)\partial b^*(k,t)\rangle$. Thus we have

$$I(k) = \operatorname{Re}\left\langle b(k,t)\frac{\partial b^*(k,t)}{\partial t}\right\rangle = \frac{\varepsilon}{16\kappa^2 k^2(k^2+\kappa^2)\pi}\int_0^\infty \mathrm{d}y |A(y)|^2 b_L(y) d_T[\omega_0(y)], \qquad (4.92)$$

where

$$A(y) = \int_0^\infty \mathrm{d}z\, R[u_s(z)](b - \mathrm{i}\kappa \tanh y)^2 \exp(-\mathrm{i}z(y-2k)/\kappa),$$

$$\omega_0(y) = 4y\kappa^2 - 8k^3 - 8k\kappa^2.$$

With the help of this correlator, we can determine the variation of a number of invariants: mass $M = \int u\,\mathrm{d}x$ and momentum $P = \int u^2\,\mathrm{d}x$. Thus

$$\frac{\mathrm{d}}{\mathrm{d}t}\langle M\rangle = \frac{4}{\pi}\int_{-\infty}^\infty I(k)\,\mathrm{d}k, \quad \frac{\mathrm{d}}{\mathrm{d}t}\langle P\rangle = \frac{8}{\pi}\int_{-\infty}^\infty k^2 I(k)\,\mathrm{d}k. \qquad (4.93)$$

For spatially uniform perturbations (4.93) is simplified to be

$$I(k) = \frac{\varepsilon^2\kappa^2}{8k^2(k^2+\kappa^2)} - |Q(k/\kappa)|^2 d_T(8k^3 + 8k\kappa^2);$$

$$Q(a) = \int_{-\infty}^{\infty} \mathrm{d}z\, R(u_s(z))(a - \mathrm{i}\tanh z)^2 \exp(-2\mathrm{i}az). \tag{4.94}$$

As a result, the spectral density of the mean energy radiated by a soliton per unit time during its motion in a homogeneous medium has the form

$$\langle p(k) \rangle = \frac{4\varepsilon^2 k^2 \kappa^2 d_T(8k^3 + 8k\kappa^2)}{\pi(k^2 + \kappa^2)} |Q(k/\kappa)|^2. \tag{4.95}$$

One can see that the main contribution to the spectral density is made by two regions in the wave number space, defined by the first two and last factors in (4.95). In particular, the dependence of total power P on k can be of a resonant character.

Let us consider, for definiteness, perturbations of the type $fR = f(t)uu_x$. It can be shown, with (4.95), that the spectral density of radiation energy is

$$\langle p(k) \rangle = \frac{\varepsilon\pi\kappa}{3^4 5^2} \Phi(k/\kappa) d_T\{8k^3(1 + \kappa^2/k^2)\},$$
$$\Phi(a) = \frac{a^6(13a^2 + 7)^2}{\sinh^2(\pi a)}. \tag{4.96}$$

If $f(t)$ is a delta-correlated function, it follows from (4.96) that the spectrum of mean power has its maximum for $k = k_{\mathrm{m}} \sim \kappa$ (Figure 4.1).

Let us proceed with the consideration of radiation of an NLS soliton in the randomly inhomogeneous medium [4.30]. The wave equation is chosen in the form

$$\begin{aligned} &\mathrm{i}q_t + q_{xx} + 2|q|^2 q = \varepsilon f(x,t) R(q); \\ &R(q) \to 0 \quad \text{for} \quad |q| \to \infty; \\ &\langle f(x,t) f(x',t') \rangle = \sigma_1^2 B_1(x - x') D_1(x - x'); \\ &\langle f(x,t) f^*(x't') \rangle = \sigma_2^2 B_2(x - x') D_2(t - t'). \end{aligned} \tag{4.97}$$

Let us evaluate now the mean spectral density of the energy of wave radiated by a soliton per unit time. This will be

$$\langle p(\lambda) \rangle = \frac{8\lambda^2}{\pi} \mathrm{Re} \left\langle b(\lambda,t) \frac{\partial b(\lambda,t)}{\partial t} \right\rangle. \tag{4.98}$$

The Jost coefficient satisfies the equation (A.36)

$$\frac{\partial b(\lambda,t)}{\partial t} = -2\mathrm{i}\lambda^2 b(\lambda,t) + \frac{\mathrm{i}\varepsilon \exp(\mathrm{i}b(t)) - 2\mathrm{i}\lambda\xi A(\lambda;\xi,\eta)}{2\eta[(\lambda - \xi)^2 + \eta^2]}. \tag{4.99}$$

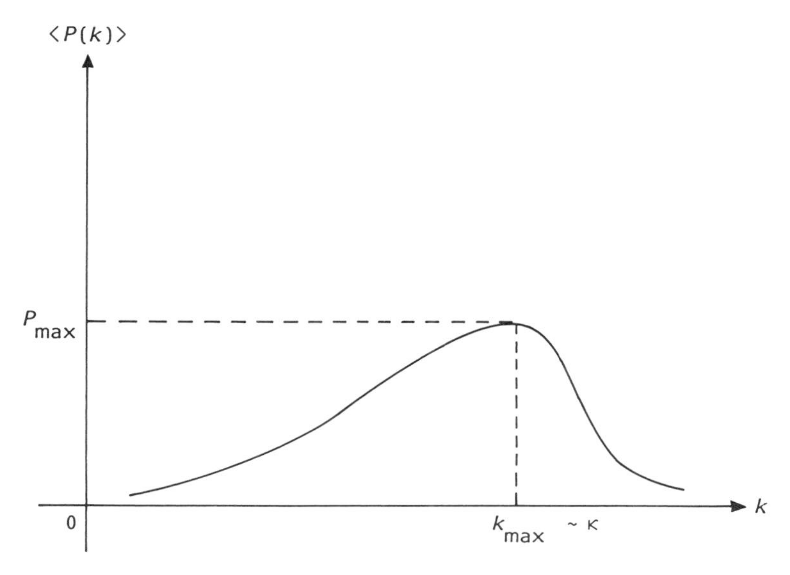

Figure 4.1. Spectrum of emitted mean power for stochastic KdV equation with perturbation $R = f(t)uu_x$ [4.30]

Algebra then leads to the following formula:

$$
\begin{aligned}
A(\lambda, \xi, \eta) &= \int \exp(-\mathrm{i}(\lambda - \xi)z/\eta)\{(\lambda - \xi - \mathrm{i}\eta \tanh z)^2 R[q_s] \exp(\mathrm{i}\Phi(z, t))\} \\
&\quad - \{\eta^2/\cosh^2 z\, R^*[q_s] \exp(\mathrm{i}\Phi(z, t)\}\,\mathrm{d}z; \\
I(\lambda) &= \mathrm{Re}\left\langle b(\lambda, t)\frac{\partial b^*(\lambda, t)}{\partial t}\right\rangle\Bigg|_{t \gg 1} = \frac{1}{8\eta^2[(\lambda - \xi)^2 + \eta^2]^2} \\
&\quad \times \int_{-\infty}^{\infty} \frac{\mathrm{d}\kappa}{\eta}\{\sigma_2^2|\varepsilon|^2\eta^4|A_1|^4 b_2(\kappa)d_2(\omega_0^{(-)}) \qquad (4.100) \\
&\quad + \sigma_2^2|\varepsilon|^2|A_2(\kappa)|^2 b_2(\kappa)d_2(\omega_0^{(+)}) + \sigma_2^2|\varepsilon|^2|A_2(\kappa)|^2 b_2(\kappa)d_2(\omega_0^{(+)}) \\
&\quad - 2\eta^2\sigma_1^2\,\mathrm{Re}[\varepsilon^2 b_1(\kappa)d_1(\omega_0^{(+)})A_1(-\kappa)A_2^*(\kappa)]\}; \\
\omega_0^{(\pm)} &= \pm 4[(\lambda - \xi)^2 + \eta^2] - 4\xi x; \\
A_1 &= \int_{-\infty}^{\infty} \mathrm{d}z \exp[-\mathrm{i}(\lambda/\eta - \xi/\eta - \kappa/2\eta)z] R \operatorname{sech}^2 z; \\
A_2 &= \int_{-\infty}^{\infty} \mathrm{d}z \exp[\mathrm{i}(\lambda/\eta + \xi/\eta - \kappa/2\eta)z] R^*(\lambda - \mathrm{i}\eta \tanh z)^2.
\end{aligned}
$$

Further we will consider the particular case of a stationary material random field. Then the formula (4.100) is essentially simplified to take the form

$$I(\lambda)=\frac{\sigma_1^2\varepsilon^2}{32\eta^2\kappa_0^2\xi^3}\{\eta^4|A_1(-\kappa_0)|^2b(-\kappa_0)+|A_2(\kappa_0)|^2b(\kappa_0) \tag{4.101}$$
$$-2\eta^2\,\mathrm{Re}[b(\kappa_0)A_1(-\kappa_0)A_2^*(\kappa_0)]\},\qquad \kappa_0=[(\lambda-\xi)^2+\eta^2]/\xi.$$

In the case of delta-correlated random processes, and with $R=q$ we have

$$\langle p(\lambda)\rangle=\frac{\pi^3\varepsilon^2\sigma_1^2\lambda^2[(\lambda-\xi)^2+\eta^2]^2}{2^6\xi^5\cosh^2[\pi(\lambda^2-\xi^2+\eta^2)/4\eta\xi]}. \tag{4.102}$$

The spectral density is seen to have two maxima (see Figure 4.2), whose width at the points $\lambda=\pm\lambda_{\mathrm{m}}$ is of order η, with

$$\lambda_{\mathrm{m}}=\sqrt{(\xi^2-\eta^2)}\sim\xi,\quad p_1\sim\varepsilon^2\sigma_1^2\xi,\quad p_2\sim\varepsilon^2\sigma_1^2\eta^4/\xi^2.$$

The total mean power ($\eta\ll\xi$) is

$$P=\int_{-\infty}^{\infty}p(\lambda)\,\mathrm{d}\lambda=4\varepsilon^2\sigma_1^2\eta\xi,$$

while for $\eta\gg\xi$,

$$P=\frac{\varepsilon^2\sigma_1^2}{\varepsilon\sqrt{2}}\xi^2(\eta/\xi)^{1/2}\exp(-\pi\eta/2\xi).$$

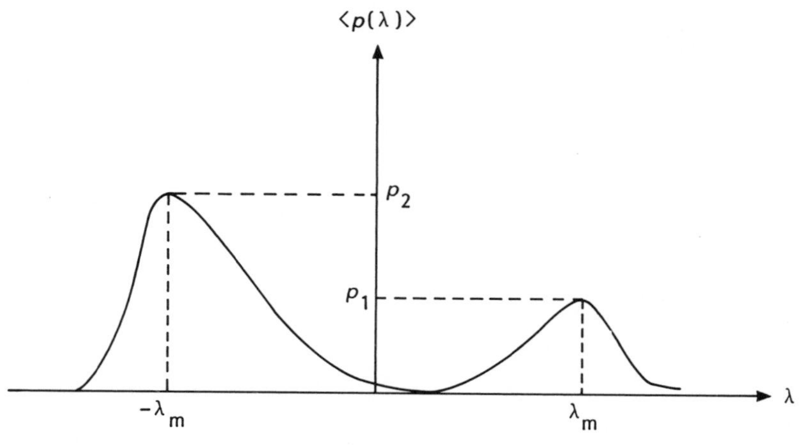

Figure 4.2. Spectral density of emitted waves for stochastic NLSE with perturbation $R=f(t)q$ [4.30]

4.5 SOLITON DYNAMICS UNDER THE ACTION OF NOISE IN NONINTEGRABLE SYSTEMS

In this section we will consider the evolution of solitons in nonintegrable systems under random perturbations. Zakharov's stochastic system and the φ^4-model seem to be the most typical problems of this type. We face the first problem, for instance, while investigating the interaction between Langmuir- and ion-noise waves in randomly inhomogeneous plasma, and between excitons and phonons in stochastic molecular crystals [4.32]. The second problem is encountered when the effect of various kinds of fluctuations on the motion of a domain interface in ferroelectrics [4.31], [4.33], [4.34] is investigated.

In this Section we will study soliton dynamics within the ψ^4-model under the action of random fields. The relevant wave equation has the form

$$mu_{tt} - mc_0^2 u_{xx} - |A|u + Bu^3 + \eta u_t = \varepsilon(x, t),$$

where $\varepsilon(x, t)$ is a random field. This equation describes the domain wall motion in a system with structural phase transitions induced by thermostat [4.33], [4.34]. Here m is the ion mass, c_0 is the sound velocity in the chain, A and B are parameters of the two-well potential, and η is a damping coefficient.

Let us consider now the case of a wall at rest,

$$u_s = u_0 \tanh(x/\sqrt{2}\xi_0), \quad \xi_0^2 = mc_0^2/|A|. \tag{4.103}$$

At low temperatures the solution can be expected to differ only slightly from (4.103). Therefore, the solution will be sought in the form

$$u = u_s + u_1, \qquad u_1 \ll u_s. \tag{4.104}$$

Substitution of (4.104) into (4.102) yields

$$u_{tt} - c_0^2 u_{1xx} + \omega_0^2[1 - \tfrac{3}{2}\operatorname{sech}^2(x/\sqrt{2}\xi_0)]u_1 = -\Gamma u_{1t} + f(x, t), \tag{4.105}$$

where

$$\Gamma = \eta/m, \qquad f = \varepsilon/m, \qquad \omega_0^2 = 2|A|/m.$$

In order to define u_1, the eigenfunctions of the operator L will be utilized. The correction u_1 is expanded over this eigenfunction basis, to give

$$u_1(x, t) = \beta_0(t)\varphi_0(x) + \beta_1(t)\varphi_1(x) + \int_{-\infty}^{\infty} dk\beta(t)\varphi(x, k). \tag{4.106}$$

Substituting (4.106) into (4.105) and taking account of the orthogonality condition, we obtain, subsequent to integration over x, an equation for $\beta_0(t)$, and hence for the soliton velocity

$$v(t) = -\frac{\sqrt{2}}{u_0}\xi_0\frac{\mathrm{d}\beta_0}{\mathrm{d}t}.$$

For low domain wall velocities, we have a stochastic equation

$$\frac{\mathrm{d}v}{\mathrm{d}t} = -\Gamma v - \frac{3}{4u_0}\int_{-\infty}^{\infty}\mathrm{d}x\,\mathrm{sech}^2[(x - vt)/\sqrt{2}\xi_0]f(x,t). \tag{4.107}$$

Since $f(x,t)$ is a Gaussian random function, it is natural to examine an equation either for the soliton velocity moments or an equation for the probability density $P(v,t)$. The solution of (4.107) has the form

$$v(t) = -\int_0^t \mathrm{d}t'\exp(-\Gamma(t-t')f_1(t') + v_0\exp(-\Gamma t));$$

$$f_1(v,t) = \frac{3}{4u_0}\int_{-\infty}^{\infty}\mathrm{d}x\,f(x,t)\mathrm{sech}^2[(x-vt)/\sqrt{2}\xi_0]. \tag{4.108}$$

Following a method put forward by Stratonovich [4.35], we obtain an equation for the probability density $P(v,t)$

$$P(v,t) = \langle\delta(v - v(t)\rangle,$$

namely

$$\frac{\partial P(v,t)}{\partial t} = \frac{\partial}{\partial v}[\Gamma vP] - \frac{\partial}{\partial v}\int\mathrm{d}\tau\int\mathrm{d}v'\langle f_1(v,t)f_1(v',t')\rangle$$
$$\times\frac{\partial}{\partial v}\left\langle\delta(v-v(t)\frac{\delta v(t)}{\delta f_1(v',\tau)}\right\rangle. \tag{4.109}$$

Equation (4.109) is closed with respect to $P(v,t)$, using the variational derivative value

$$\frac{\delta f_1(v,t)}{\delta f_1(v',t')} = \delta(v-v')\delta(t-t'), \tag{4.110}$$

and the solution of (4.108). The resulting equation is

$$\frac{\partial P}{\partial t} = \frac{\partial}{\partial v}[\Gamma vP] - \frac{\partial}{\partial v}D(t)\frac{\partial P}{\partial t}, \tag{4.111}$$

wherein a time-dependent diffusion coefficient is introduced,

$$D(t) = \int_0^t d\tau \exp(-\Gamma(t-t'))\langle f(v,t) f(v,\tau)\rangle. \qquad (4.112)$$

The initial condition for this equation is

$$P(v,t)|_{t=0} = \delta(v - v_0).$$

Now we will consider δ-correlated random forces,

$$K(x-y, t-\tau) = 2\sigma_\varepsilon^2(T)\delta(x-y)\delta(t-\tau), \qquad (4.113)$$

where T is temperature. From (4.111), a stationary solution is readily derived as

$$P_{st}(v) = \exp(-\Gamma v/2D)/\sqrt{(2\pi D)}, \qquad (4.114)$$

where the kink diffusion coefficient is

$$D = \frac{3\sqrt{2}\xi\sigma_\varepsilon^2(T)}{4u_0^2 m^2}. \qquad (4.115)$$

Identification of (4.114) with the thermal distribution leads to

$$\sigma_\varepsilon^2 = 2\eta T.$$

A nonstationary distribution of the soliton (kink) velocity is also defined to be

$$P(v,t) = \sum_{n=0}^{\infty} \sqrt{\left(\frac{1}{2\pi D}\right)} \frac{1}{2^n n!} \exp[-n\Gamma t]$$
$$-\frac{\Gamma}{2D}(v^2 - v_0^2) H_n\left[\left(\frac{\Gamma}{2D}\right)^{1/2} v_0\right] H_n\left[\left(\frac{\Gamma}{2D}\right)^{1/2} v\right]. \qquad (4.116)$$

Here H_n is the Hermite polynomial. Equation (4.116) allows the construction of moments of arbitrary order, for instance

$$\langle v^2\rangle = \langle v_0^2\rangle[1 + (v_0^2/\langle v_{st}^2\rangle - 1)\exp(-2\Gamma t)];$$
$$\langle v_{st}^2\rangle = D/\Gamma. \qquad (4.117)$$

Given that the correlation length l_ε and time τ_ε are finite, the correlation function can be written in the following form:

$$K(x-y, t-\tau) = 2\sigma_\varepsilon^2 U(x-y)V(t-\tau),$$

$$U(x-y)=\exp(-|x-y|/l_\varepsilon)/2l_\varepsilon,$$
$$V(t-\tau)=\exp(-|t-\tau|/\tau_\varepsilon).$$

Substituting these expressions into (4.112), we come to the time-dependent diffusion coefficient as

$$D_{\tau_\varepsilon}(t)=6D_\alpha[\psi'(\alpha)-1/\alpha-1/2\alpha^2]\frac{1-\exp[-(1+\Gamma\tau_\varepsilon)t/\tau_\varepsilon]}{1+\Gamma\tau_\varepsilon}, \tag{4.118}$$

where $\psi'(\alpha)$ is the first derivative of ψ ($\alpha=\sigma_0/\sqrt{2l_\alpha}$). It is easy to see that for $l_\varepsilon, \tau_\varepsilon \to 0$ this result reduces to (4.115). From comparison of the two expressions we can infer that: first, the expansion over $t-\tau$ is equivalent to that over the parameter $\Gamma\tau_\varepsilon$; secondly, the finiteness of the correlation length leads to no new effects except for renormalization of the diffusion coefficient; and thirdly, the finite correlation length is essential only for $\tau \leqslant \tau_\varepsilon$.

It is to be noted that the relation (4.116) allows the evolution of all the moments without resort to the solution of the Fokker–Planck equation. For instance, for the mean of v^2 we have

$$\langle v^2\rangle=\frac{D_{\tau_\varepsilon}(\infty)}{\Gamma(1-\Gamma\tau_\varepsilon)}[1-\exp(-2\Gamma t)-\Gamma\tau_\varepsilon(1+\exp(-2\Gamma t)$$
$$-2\exp(-(1+\Gamma\tau_\varepsilon)t/t_s)].$$

In the limit $t\to\infty$ this to

$$\langle v^2\rangle_{st}=\frac{D_{\tau_\varepsilon}(\infty)}{\Gamma},$$

and this is reproduced from the Fokker–Planck equation.

4.6 NUMERICAL SIMULATION OF SOLITON DYNAMICS UNDER RANDOM PERTURBATIONS

While investigating the dynamics of solitons and other types of nonlinear waves in randomly inhomogeneous media, of particular interest is the development of methods that are not associated with the application of perturbation theory. In this connection numerical simulation of the relevant problems is useful. Unfortunately, there are a few papers on this topic [4.13], [4.36], [4.38], [4.44]–[4.46]. This is probably due to the fact that, from the point of view of computational difficulties, the progression to stochastic problems is at least analogous to the progression to higher dimensions.

Here we consider the basic results of numerical simulations of soliton evolution. Let us examine the KdV equation with random parameters, describing the interaction between a soliton and a fluctuating wave [4.38]:

$$u_t + 6uu_x + u_{xxx} = \varepsilon(x)u, \tag{4.119}$$

where $\varepsilon(x)$ is a random Gaussian function, $\langle\varepsilon\rangle = 0$, and $\langle\varepsilon(x)\varepsilon(y)\rangle = B(x-y, l_\varepsilon)$; here l_ε is a correlation length. For $l_\varepsilon \to 0$, $B(x) \to 2\sigma_\varepsilon^2\delta(x)$. The equation (4.119) describes the parametric interaction of a soliton with synchronously moving fluctuations. We will analyze first of all an analytic description of the soliton in the early stages of its motion, in terms of the IST. The solution is sought in the form

$$u(x,t) = -2\kappa^2(t)\operatorname{sech}^2 z; \qquad z = \kappa(\kappa - \xi(t)). \tag{4.120}$$

In the adiabatic approximation (A.94) we find for the soliton amplitude ξ and velocity $d\xi/dt$ the following expressions:

$$\frac{d\kappa}{dt} = 4\kappa^2 \int_{-\infty}^{\infty} \varepsilon(z/\kappa + \xi)\operatorname{sech}^4 z\, dz, \tag{4.121}$$

$$\frac{d\xi}{dt} = 4\kappa^2 + \frac{1}{4\kappa^2}\int_{-\infty}^{\infty} \operatorname{sech}^2 z(z + \tfrac{1}{2}\sinh 2z)\varepsilon(z/\kappa + \xi)\, dz. \tag{4.122}$$

For small $\varepsilon \ll 1$ and $t \ll \langle\varepsilon^2\rangle$ we have

$$\langle\kappa^2\rangle = 4\int_0^t \langle\kappa^2(t)\rangle dt + O(\varepsilon^2) \approx \frac{2\kappa_0^2}{\sigma_\varepsilon^2}[\exp(2\sigma_\varepsilon^2 t) - 1]. \tag{4.123}$$

The soliton amplitude is seen to grow. The evolution at times $t \sim \varepsilon^2$ can be analyzed by the mean field method (see Section 4.1). The calculation shows that the growth behavior for $t > t_*$ is modified by saturation effects and the amplitude becomes stationary in time [4.40], [4.39].

Now we will report results of numerical simulation [4.38]. The computer simulation scheme ran as follows. We treated a regularized KdV equation

$$u_t + u_x + uu_x - u_{xxt} = \varepsilon(x)u. \tag{4.124}$$

For small amplitudes this is known to reproduce KdV solutions [4.42]. Equation (4.124) was solved using an implicit difference scheme suggested in Ref. [4.41]. The initial condition was chosen in the form

$$u(x,t) = 3c\,\mathrm{sech}^2(kx - \omega t + \delta), \quad c = \frac{a^2}{1-a^2}, \quad k = \frac{a}{2}$$

$$\omega = \frac{k}{1-a^2}, \tag{4.125}$$

where $\delta = -7$, $a^2 = \frac{1}{16}$. The ranges of variation of x and t were $0 < x < 200$, $0 < t < 114$, Δx is a step over x, Δt is a step over t, $\Delta t = 0.2$. Twenty different realizations of the random variable $\varepsilon(x)$ were run, a uniform distribution being chosen, and $-0.00769 < \varepsilon(x) < 0.00769$. The mean field was computed from 20 realizations; increase in the number of realizations did not change the final results. The results of simulation of the mean field $\langle u \rangle$ are shown in Figure 4.3. In the initial stage (up to times $t \sim 30.4$) a stochastic acceleration is observed (Figure 4.3) indicating the analogy between a soliton and a particle and agreeing with the results of perturbation theory analysis. However, from time ~ 60 the soliton amplitude decreases and from time $t \sim 114$ a second soliton is formed. Then the second stage of the growth of the leading soliton begins. It is open to question whether solitons can be multiply generated by a parametrically acting random field in this way. This issue needs further analysis.

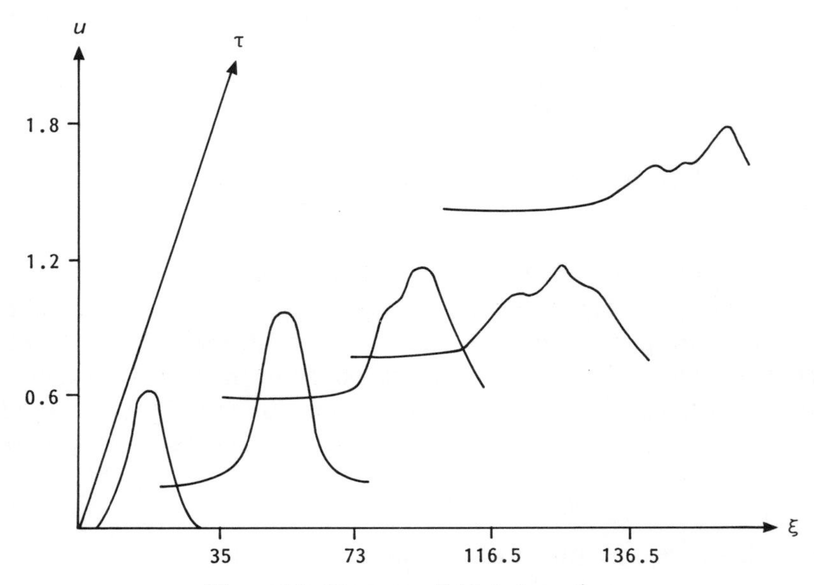

Figure 4.3. The mean field $\langle u \rangle$ vs. time

The second case is associated with the propagation of an SG soliton in a randomly inhomogeneous medium. The corresponding stochastic SG equation is

$$u_{tt} - u_{xx} + \sin u = \varepsilon(x)u_{xx}. \tag{4.126}$$

Let us report the results of numerical simulation [4.28], [4.36]. The initial condition was chosen as an SG soliton. In dimensionless variables, x and t vary over the ranges $0 < t < 15$, $-50 < x < 50$ ($k = 0.05$ and $h = 0.1$ are discretization steps over x and t). With the help of a random number generator 15 realizations of the random function were run over 1000 points for each implementation. Then with a scheme suggested in Ref. [4.43] a random process with an exponential correlation function was built up. The correlation length was $l_\varepsilon = 6$, $\sigma_\varepsilon^2 = 0.5$, $d_s \sim l_\varepsilon$, where d_s was the soliton width. The results obtained from the numerical experiment generally confirm the analytic estimates. The energy of waves radiated by a soliton is seen to be rather low for $v \ll 1$ and to increase sharply for $v \to 1$. The soliton loses energy upon radiating, which decreases the velocity. In the case when the correlation length $l_s \sim d_s$ is of the order of the soliton width, small-amplitude breathers are formed. The evolution of solitons and breathers under strong random fields was considered in Refs. [4.46], [4.47].

To investigate the problem of soliton propagation in optical waveguides for long distances (more than 1000 km), numerical modeling of NLS equation with linear and nonlinear stochastic perturbations seems to be interesting. These perturbations describe random variations of the optical fiber parameters. The consideration of this problem has just now been started (see Ref. [4.44]). The nonlinear tunneling in NLS and SG with random parameters numerically investigated in Refs. [4.48], [4.49].

4.7 BREATHERS IN A RANDOM FIELD

As is known, together with solitons there are other nonlinear modes of integrable systems-breathers and N-soliton complexes. Their dynamics in a random field may have new features compared with that of single solitons. For example, stochastic breather decay into a soliton–antisoliton pair, stochastic resonance upon the interaction of soliton complexes with noise and other processes belong to these new phenomena.

Let us consider the interaction of a SG breather with a random field. The corresponding stochastic SG equation is

$$u_{tt} - u_{xx} + \sin u = \varepsilon f(x, t)R(u), \tag{4.127}$$

where $f(x, t)$ is the random field. Here there are two limiting cases convenient for analysis: breathers of small and large amplitude. A small-amplitude breather bears a close resemblance to an NLS soliton in shape and is described by the following expression:

$$u_{\mathrm{B}} = 4\mu \sin\left[(1-\mu^2/2)(t-vx)/\sqrt{(1-v^2)}\right] \operatorname{sech}(\mu(x-\xi(t))/\sqrt{(1-v^2)}), \tag{4.128}$$

where μ is the breather amplitude.

A breather of large amplitude has the form

$$u_{\mathrm{B}}(x, t) = 4\tan^{-1}\{\sinh[r(t)/2]/\cosh x\}, \tag{4.129}$$

or

$$u_{\mathrm{B}} = 4\tan^{-1}\{T(t)/\cosh x\}, \quad \frac{\mathrm{d}T}{\mathrm{d}t} = \pm\sqrt{(1-\zeta^2 T^2)}, \quad \zeta = \pi/2 - \mu \ll 1. \tag{4.130}$$

Choose $R(u) = \sin u$.

Initially we will investigate the behavior of a large-amplitude breather (see Sect. 1.3). Let us define a probability distribution for the soliton–antisoliton relative distance $P_t(r, \dot{r})$. Applying a standard procedure, we obtain the Fokker–Plank equation for P_t,

$$\frac{\partial P_t(v, z_0)}{\partial t} + \{H, P_t\} - \Gamma\frac{\partial}{\partial v}(vP_t) = \sigma_f^2\frac{\partial^2 P_t}{\partial v^2}; \quad \{H, P_t\} = v\frac{\partial P_t}{\partial z_0} - \frac{\partial H}{\partial z_0}\frac{\partial P_t}{\partial v}. \tag{4.131}$$

Hence we see that a stationary distribution looks like the Gibbs distribution,

$$P_{\mathrm{st}}(v, z_0) = \mathrm{const}\exp\left\{-\frac{\Gamma}{\sigma_f^2}(v^2/2 + u(z_0))\right\}, \tag{4.132}$$

Integration over the relative velocities yields the Boltzmann distribution in the coordinates,

$$\begin{gathered} P_\infty(z_0) = \mathrm{const}\exp(-\Gamma\, U(z_0)/\sigma_f^2), \\ U(z_0) = \exp(-|z_0|) \end{gathered} \tag{3.133}$$

Here a quasi-particle method has been applied. It is useful for analyzing systems that are not integrable when the perturbation is switched off. For the case of systems close to integrable ones, the most convenient is the method based on, the IST. Here we report the results of the IST-based analysis [4.46], [4.47]. To begin with we will use expression (4.130) for a large-amplitude breather. An equation for the breather amplitude was obtained by perturbation theory for breathers, in the form

$$\frac{d\zeta^2}{dt} = -\frac{\pi\varepsilon}{2}(f(t)\sqrt{[(1-\zeta^2T^2)(1+T^2)]}. \tag{4.134}$$

From this one can derive the following Fokker–Planck equation for the distribution function $P(z, T; t)$, $z = \zeta^2$:

$$\frac{\partial P}{\partial t} = \pm\frac{\partial P}{\partial T}[(1-zT^2)^{1/2}P] + \frac{\pi^2\varepsilon^2}{16}T^2(1+T^2)\frac{\partial P}{\partial z} + \frac{\pi^2\varepsilon^2}{8}\frac{\partial^2}{\partial z^2}[(1-zT^2)(1+T^2)^{-1}P]. \tag{4.135}$$

The initial condition is chosen in the form

$$P(z, T; t=0) = \frac{\zeta_0}{\pi}(1-\zeta_0^2T^2)^{-1/2}\delta(z-\zeta_0^2), \qquad \zeta_0^2 \ll 1. \tag{4.136}$$

It describes the initial state with a given value of ζ_0^2; the phase is considered to be uniformly distributed. A diffusion time is estimated as the time needed for $z = 0$ to be reached. The Fokker–Planck equation (4.135) is hardly solvable. However, for $t_d \ll \zeta_0^{-1}$, $\zeta_0^3 \ll \varepsilon^2$ one can obtain an approximate solution [4.46],

$$P(z, T; t) = \frac{\zeta_0}{\varepsilon}\left[\frac{2}{\pi^5}(1+T^2)(1-\zeta_0^2T^2)^{-1}(1-zT^2)^{-1}\right]^{1/2} \times t^{-1/2}\exp[-(8/\pi^2\varepsilon^2T^2t)(2-T^2)(z+\zeta_0^2))] \times \cosh[(16/\pi^2\varepsilon^2T^4t)(1+T^2)\sqrt{[(1-zT^2)(1-\zeta_0^2T^2)]}]. \tag{4.137}$$

For $T\to 0$ this turns into a solution of the diffusion equation,

$$P(z, T; t) = \frac{\zeta_0}{\varepsilon}\sqrt{\left(\frac{2}{\pi^5 t}\right)}\exp[-2(z-\zeta_0^2)^2/(\pi^2\varepsilon^2 t)]. \tag{4.138}$$

In a similar way analysis of the dynamics of a small-amplitude breather in an external random can be performed. The initial equation is [4.27].

$$u_{tt} - u_{xx} + \sin u = \varepsilon f(t) - \alpha u_t. \tag{4.139}$$

In the adiabatic approximation equations for the breather amplitude and phase have the form (A.54–57)

$$\frac{d\gamma}{dt}=\frac{\pi f(t)\sin\theta}{4\cos\gamma(1+\tan^2\gamma\cos^2\theta)^{1/2}}-\frac{\alpha\tan\gamma\sin^2\theta}{1+\tan^2\gamma\cos^2\theta}\times\left[1+\frac{\tan^{-1}\gamma\sinh^{-1}(\tan\gamma\cos\theta)}{\cos\theta(1+\tan^2\gamma\cos^2\theta)^{1/2}}\right]; \tag{4.140}$$

$$\frac{d\theta}{dt}=\cos\gamma-\pi\gamma\left[\frac{\cos\theta}{\cos^2\gamma(1+\tan^2\gamma\cos^2\theta)^{1/2}}-\tan^{-1}\gamma\sinh^{-1}(\tan\gamma\cos\theta)\right]-\alpha\frac{\cos\theta\sin\theta}{\cos^2\theta(1+\tan^2\gamma\cos^2\theta)}-\alpha\frac{\tan\gamma\sinh^{-1}(\tan\gamma\cos\theta)\sin^3\theta}{(1+\tan^2\gamma\cos^2\theta)^{3/2}}. \tag{4.141}$$

For a small-amplitude breather ($\gamma \ll 1$) this set of equations simplifies to

$$\frac{d\gamma}{dt}=\frac{\varepsilon\pi f(t)}{4}\sin\theta; \tag{4.142}$$

$$\frac{d\theta}{dt}=1-\frac{\alpha}{2}\sin\theta. \tag{4.143}$$

In this case the breather velocity is constant. The system is seen to be decoupled and in a γ-linear approximation, the random perturbation does not affect the breather phase. Let us evaluate first of all the breather phase. A solution of (4.143) has the form

$$t=\frac{1}{(1-B^2)}\tan^{-1}\left[\frac{\tan\theta-B}{\sqrt{(1-B^2)}}\right],\qquad B=\alpha/2.$$

Having substituted this result into the equation (4.142) for the amplitude, we obtain a closed equation for γ,

$$\frac{d\gamma}{dt}=\frac{\pi\varepsilon f(t)}{4}\sin\alpha\tan^{-1}(\tan t+B), \tag{4.144}$$

from which we have

$$\gamma=\frac{\varepsilon\pi}{4}\int_0^t\frac{f(t)(\tan t+B)\,dt}{\sqrt{(1+\tan^2 t+2B\tan t)}}.$$

Evaluation of the integral yields.

$$\langle \gamma^2 \rangle = \frac{\pi^2 \varepsilon^2 \sigma_f^2}{16}\left[t - \frac{2}{\alpha}\ln\left|1 + \frac{\alpha}{2}\sin 2t\right| + \tan^{-1}\left|\frac{\alpha}{2} + \tan t\right| + \tan^{-1}\frac{\alpha}{2}\right]. \tag{5.145}$$

For $\alpha \to 0$ we have

$$\langle \gamma^2 \rangle = \frac{\pi^2 \varepsilon^2 \sigma_f^2}{8}\left(t - \frac{\sin 2t}{t}\right),$$

which is consistent, as expected, with the behavior of the NLS soliton in a random field (see Section 4.3).

In conclusion we well consider the radiation of waves by a breather in a random field [4.27], [4.37]. The investigation of a small-amplitude limit is of interest since in this case the calculations can be performed exactly, and second, small-amplitude breathers are readily excited by external forces [4.49] and unlike SG solitons they can in an essential way contribute to the noise of different devices (using the LJJ for example).

An equation for the Jost coefficient $b(\lambda, t)$, important for calculations of the emitted energy, has the form

$$\frac{\mathrm{d}b}{\mathrm{d}t} = -2\mathrm{i}\omega(\lambda)b - \frac{\mathrm{i}\varepsilon}{4}\int_{-\infty}^{\infty} R_{\mathrm{B}}[u]\exp(-2\mathrm{i}k(\lambda)x)\,\mathrm{d}x, \tag{4.146}$$

where $R(u_{\mathrm{B}})$ is the perturbation, and a breather solution of the unperturbed SG equation is (4.128). The perturbation of the SG equation is chosen in the form

$$R(u) = f(x)\sin u. \tag{4.147}$$

This type of perturbation occurs while studying a Josephson transition with random grating. In the following an exponential model of the correlator is used. With this expression, we obtain the following equation for $b(\lambda, t)$:

$$\frac{\mathrm{d}b}{\mathrm{d}t} = -2\mathrm{i}\omega(\lambda)b - \mathrm{i}\varepsilon\mu\int_{-\infty}^{\infty} \exp(-2\mathrm{i}k(\lambda)x\frac{\sin[(t - vx)/\gamma]\,\mathrm{d}x}{\cosh[\mu(x - vt)/\gamma]}. \tag{4.148}$$

Mean radiated power is calculated from the formula

$$\frac{\mathrm{d}\langle E\rangle}{\mathrm{d}t} = \frac{2}{\pi}\int\frac{\mathrm{d}\lambda}{\lambda}\omega(\lambda)\mathrm{Re}\left\{\left\langle b^*\frac{\partial b}{\mathrm{d}t}\right\rangle\right\}. \tag{4.149}$$

After some algebra the mean radiated power is found to be

$$\frac{d\langle E\rangle}{dt} = \varepsilon^2 \alpha \gamma^2 \int_1^\infty d\omega P(\omega), \tag{4.150}$$

where

$$P(\omega) = \frac{1}{k}\left[\frac{1}{(1+\kappa_1^2 l_\varepsilon^2)\cosh^2[\pi(\omega\gamma+1)/2v\mu]} + \frac{1}{(1+\kappa_2^2 l_\varepsilon^2)\cosh^2[\pi(\omega\gamma-1)/2v\mu]}\right]$$

$$k = \sqrt{(\omega^2-1)}, \qquad \kappa = 2k - \frac{2\omega \pm \gamma}{v}.$$

In deriving (4.150) we have proceeded from, integration over a spectral parameter λ to integration over frequencies ω and used the following relation:

$$\frac{dk}{k} = \frac{d\lambda}{\lambda}$$

The qualitative behavior of $P(\omega)$ shows that at the frequency $\omega = 1/\gamma$ there is a maximum due to *stochastic resonance* between the random perturbation and the inner breather oscillations. Let us inspect expression (4.150) in more detail. Let $\pi\gamma/l_\varepsilon\mu \gg 1$, i.e. fluctuations are small-scale. Then

$$\frac{d\langle E\rangle}{dt} = \varepsilon^2\gamma^2\alpha\left\{\frac{\mu}{\pi\gamma}(2-\exp[-(1-\gamma)]) - \sqrt{\left(\frac{\pi^3\gamma}{v\mu}\right)}\frac{\exp[-\pi(1+\gamma)/(v\mu)]}{1+(1+\gamma)^2 l_\varepsilon^2/v^2}\right\}.$$

We can see that unlike soliton radiation, the radiation of a breather is not exponentially weak in velocity and amplitude. It should be noted that a breather of zero velocity radiates as well. The radiated power is

$$\frac{d\langle E\rangle}{dt} = \frac{\varepsilon\alpha\mu}{\pi}.$$

Similarly one can consider radiation under the action of time-fluctuations, when $R = \varepsilon f(t)\sin u$ [4.27]. Analogously one can calculate the spectral density of the breather emission is an external random field [4.47]. Let us define the spectral density of radiated energy by

$$W(\chi) = W(k)\frac{dk}{d\chi}, \qquad W = \frac{dE}{dt}.$$

After calculation we obtain

$$\langle W(\chi)\rangle = \frac{\pi}{2^{7/2}}\varepsilon^2(\chi-1)^{-1/2}\,\mathrm{sech}^2(\pi\sqrt{(\chi-1)}/\sqrt{2}\mu);$$

Here $\chi - 1$ plays the role of the $\varphi - \mu^2/2$ parameter. This expression is singular at $\chi - 1$. This singularity is smoothed if we take into account the influence of dissipation. The total energy converges and is equal to

$$\langle W\rangle = \int_1^\infty \langle W(\chi)\rangle\,\mathrm{d}\chi = \frac{\varepsilon^2\mu}{4}.$$

It is clear that in this case the exponential smallness of breather radiation is absent also: breather radiation is algebraically only small.

5

DYNAMICAL CHAOS OF SOLITONS AND BREATHERS IN INHOMOGENEOUS MEDIA

5.1 BROWNIAN MOTION OF THE SG SOLITON IN THE FIELD OF TWO WAVES

Let us consider the motion of a soliton in the field of two waves. The initial wave equation is

$$\varphi_{tt} - c_0^2 \varphi_{xx} + \omega_0^2 \sin\varphi + \Gamma\varphi_t = \varepsilon f(x,t) R(\varphi), \tag{5.1}$$

where c_0 is the cutoff velocity, ω_0 is the characteristic frequency, Γ is damping, $f(x,t)$ is the periodic external force, and $R(\varphi)$ is the perturbation operator ($R(\varphi) = 1, \varphi_{xx}$ etc.). As an example we will study the motion of a magnetic soliton in a quasi-one-dimensional ferromagnet exhibiting 'easy plane' anisotropy in the field of two sound waves [5.1]. In this case (5.1) will have the form

$$\varphi_{tt} - c_0^2 \varphi_{xx}(1 + \varepsilon\xi(x,t)) + \omega_0^2 \sin\varphi + \Gamma\varphi_t = 0. \tag{5.2}$$

Here $c_0^2 = 2AIS^2a^2, \omega_0^2 = 2g\mu\beta H_0^x S$, A is the anisotropy constant, I is the exchange interaction, S is spin, a is the atomic spacing, H_0^x is the constant magnetic field. The sound field is represented by $\xi(x,t)$; $\xi(x,t) = \alpha/(I \cdot U_x(x,t))$, where α is the coefficient of spin–phonon coupling, U is the field of lattice displacement.

Let a single sound wave travel along a ferromagnetic crystal,

$$U(x,t) = U_0 \sin(kx - \omega t).$$

Using standard dimensionless variables $t \to t\omega_0, x \to x/\alpha, d = C_0/\omega_0$, $\alpha = \Gamma/\omega_0$, $\varepsilon = \alpha U_0 \omega_0/(Ic_0), k_1 = kc_0/\omega_0, \omega_1 = \omega/\omega_0$, we obtain the equation

$$\varphi_{tt} - \varphi_{xx} + \sin\varphi + \alpha\varphi_t = -\varepsilon \sin(k_1 t - \omega_1 x)\varphi_{xx}. \tag{5.3}$$

Note that renormalization is not needed, since the term $f(x,t)R(\varphi)$ descreases for $|x| \to \infty$, i.e. the perturbation is localized on the soliton. Due to the smallness of α and β we can apply perturbation theory for SG solitons (see Appendix 2). Thus we obtain the following set of equations for soliton velocity $V(t)$ and soliton center coordinate $\xi(t)$:

$$\frac{dv}{dt} = \frac{\varepsilon}{4}(1 - v^2)^{3/2} \frac{\pi\omega_1^2 \cos(k_1 t_1 - \omega_1 \zeta)}{\sinh[(\pi\omega_1(1 - v^2/2)^{1/2})/2]} - \alpha v, \tag{5.4}$$

$$\frac{d\zeta}{dt} = v(t) + O(v\varepsilon). \tag{5.5}$$

For low soliton velocities $v^2 \ll 1$, the soliton center coordinate is expressed as follows:

$$\frac{d^2\zeta}{dt^2} + \alpha\frac{d\zeta}{dt} = \varepsilon A \cos(k_1 t - \omega_1 \zeta), \tag{5.6}$$

$$A = \omega_1^2/(4\sinh(\pi\omega/2)).$$

In its appearance this equation coincides with the equation of motion of a charged particle in the wave field.

Examine now the case when $\alpha = 0$. There are two types of soliton motion: trapped, corresponding to oscillations with the period

$$T = 2\pi/(\varepsilon A), \tag{5.7}$$

and propagating, occurring under the condition that

$$H > H_s,$$

where H_c is the value of the Hamiltonian in (5.6) on the separatrix ($\alpha = 0$).

Inserting a second wave into (5.3) destroys the separatrix that separates trapped and propagating motions, leading to the occurrence of a stochastic layer in the vicinity of the separatrix. Given initial conditions such that the soliton finds itself inside the stochastic layer, the motion has a mixing nature to be characterized by the exponential decrease of phase correlations,

$$\langle \exp(\mathrm{i}\varphi)\exp(\mathrm{i}\varphi')\rangle \simeq \exp(-\ln K), \qquad K \gg 1, \tag{5.8}$$

where K is a parameter of stochasticity as defined in [5.2].

By virtue of the formal analogy between (5.6) and the motion of a particle in the wave field, we can apply the results of Ref. [5.2] to estimate a criterion for stochasticity K and the width of the stochastic layer near the separatrix $\bar{\omega}$ (Figure 5.1). So we have

$$K^2 = \frac{\varepsilon_1}{16\pi}\exp\left[\frac{\pi}{\tau\omega}\right],$$
$$\varepsilon_1 = \frac{A_1}{A}, \tag{5.9}$$
$$\bar{\omega} = \frac{\pi}{T\ln(16\pi/\varepsilon_1)}.$$

Note that these estimates have been obtained with no dissipation ($\alpha = 0$) in the system. Dissipation can be taken into account in the framework of Melnikov's method [5.3], allowing the evaluation of separatrix

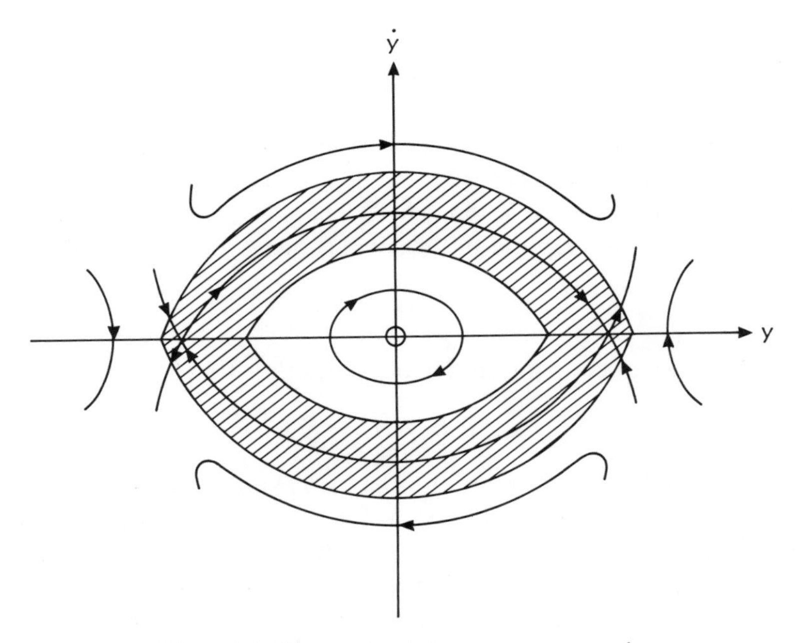

Figure 5.1. The stochastic layer near separatrix

splitting under perturbation. For the external wave ($R = 1$) the boundary value of damping α, at which stochasticity is still feasible (Ref. [5.4]), is calculated to have the form

$$\alpha^* = \left| \frac{\pi\omega_0^2[\cosh^{-1}(\omega\pi) - \sinh^{-1}(\omega_0\pi)]}{A_1(1 + \pi\omega_0/2)} \right|. \tag{5.10}$$

Here A_1 is the amplitude of the travelling wave, $\omega_0 = \omega/\sqrt{[(\pi/4)A_1K]}$ For $\alpha > \alpha^*$ stochasticity is not feasible. To obtain this estimate the wave field has been renormalized.

And now we consider the soliton motion in the field of a wavepacket [5.1].

The wave equation in this case is

$$\varphi_{tt} - \varphi_{xx} + \sin\varphi = \sum(E_k\, \mathrm{e}^{\mathrm{i}(kx - \omega(k)t)} + \text{c.c.}). \tag{5.11}$$

Acting as previously, we obtain the equation for the soliton center coordinate as

$$\frac{\mathrm{d}^2\xi}{\mathrm{d}t^2} = \frac{\varepsilon}{4}\sum_k \bar{A}_k E_k \mathrm{e}^{\mathrm{i}(k\xi - \omega(k)t)},$$

$$\bar{A}_k = \pi/[(k + \omega^2)\cosh(\pi k/2)]. \tag{5.12}$$

Following Ref. [5.5] we will define the regions of parameters wherein the soliton motion becomes stochastic. To begin with, we consider the case when a soliton is trapped by any packet harmonic. This forms an ergodic layer for the soliton. Its frequency band is of the order of

$$\tau_k^{-1} = \tfrac{1}{2}(\varepsilon A_k E_k)^{1/2}$$

To evaluate the stochastic region in which the soliton moves in a random manner, we will impose the condition of coalescence of ergodic layers. This gives the following criterion for stochasticity:

$$\tau_k\Omega_k = 2\Omega_k(\varepsilon A_k E_k)^{1/2} \ll 1, \tag{5.13}$$

where frequency $\Omega_k = \omega(k + \Delta k) - \omega(k)$ is a slowly varying function of K. The soliton velocity distribution function $p(v, t)$ in the stochastic region has the form

$$\frac{\partial p(v,t)}{\partial t} = 2\frac{\partial}{\partial v}\sum_k (\varepsilon A_k E_k/4)^{1/2}\frac{\tau_r}{1 + \omega^2(k)\tau_r^2}\frac{\partial p(v,t)}{\partial v}, \tag{5.14}$$

where τ_r is a time for phase correlation splitting, related to soliton and packet parameters as follows:

$$\tau_r^{-1} = \tau^{-1} \ln(1/(\tau\Omega)), \qquad \tau = \langle \tau k \rangle.$$

As is seen from (5.1.14), the mean soliton energy increases with time, i.e. the soliton is stochastically accelerated.

5.2 CHAOS OF AN ENVELOPE SOLITON IN THE PERIODIC FIELD

In this section we will study the dynamics of an NLS soliton in the low-frequency wave field [5.6]. A set of equations for this problem is

$$\begin{aligned} i\psi_t + \psi_{xx} - n\psi &= 0, \\ n_{tt} - n_{xx} &= (|\psi|^2)_{xx}. \end{aligned} \tag{5.15}$$

Here ψ is a complex envelope of the high-frequency field, n is the low-frequency field, $n = n_0 + n(\tilde{x}, t)$, and $\tilde{n}(x, t)$ describes the medium inhomogeneity. These equations describe the interaction between langmuir and ion-sound waves in nonisothermal plasma, the interaction between excitons and phonons in molecular chains, etc.

What is the dynamics of the soliton for $n \neq$ constant, $n = \tilde{n}(x, t) - |\psi|^2/(1 - v^2)$? When $\tilde{n}(x, t)$ is a smooth function of x and t as compared with soliton dimensions, the problem can be treated using asymptotic methods. Following direct methods (see Section 1.8), the solution of (5.15) is sought in the form

$$\varphi = \left[\Phi^{(0)}(\zeta, \rho, \tau) + \sum_{n=1} \varepsilon^n \Phi^n(\zeta, \rho, \tau) \right] \exp\{i(v/2) + \varphi)\}, \tag{5.16}$$

$$n = n^{(0)}(\zeta, \rho, \tau) + \sum_{n=1} \varepsilon^n n^{(n)}(\zeta, \rho, \tau), \tag{5.17}$$

where

$$\begin{aligned} &\Phi^{(0)} = [2\lambda^2(1 - v^2)]^{1/2} \cosh^{-1} \lambda\zeta, \\ &\zeta = x - vt, \qquad \varphi = \omega t, \\ &\lambda^2 = \omega + \tilde{n} - v^2/4, \quad \rho = \varepsilon x, \quad \tau = \varepsilon t, \\ &\varepsilon \ll 1, \qquad \varepsilon \sim l_s/l_{\tilde{n}}. \end{aligned} \tag{5.18}$$

From the condition of orthogonality (1.75) one readily derives equations for the soliton parameters:

$$\frac{\mathrm{d}}{\mathrm{d}t}\lambda(1-v^2)=0,$$

$$\lambda(1-v^2)\left(\frac{\mathrm{d}v}{\mathrm{d}t}+2\frac{\partial\tilde{n}}{\partial x}\right)+\frac{8}{3}\frac{\mathrm{d}}{\mathrm{d}t}(v\lambda^3)=0. \tag{5.19}$$

This set of equations can be written as a single equation for the soliton center coordinate $x(t)=\int_0^t v(t')\,\mathrm{d}t'$:

$$\frac{\mathrm{d}^2x}{\mathrm{d}t^2}=-\frac{2\tilde{n}_x}{1+\frac{8}{3}c^2\dfrac{1+5\dot{x}^2}{(1-\dot{x}^2)^4}}, \tag{5.20}$$

where c is a constant indicating the number of 'quanta'. For $|\dot{x}|\ll 1$ this equation coincides with the equation of motion of a classical particle in the external field. As the analysis Ref. [5.2] shows, for stochastization of the particle motion the field spectrum $n(x,t)$ should have additional harmonics apart from the main wave.

The equation of soliton motion in the field of the standing wave was numerically integrated on the computer from the following [5.6]:

$$\frac{\mathrm{d}^2x}{\mathrm{d}t^2}=E\sin x\sin t \tag{5.21}$$

Here $E=\frac{3}{4}N_0/c^2$, and N_0 is the amplitude of the sound wave. The motion was found to be stochastic, representing the Poincaré sections. The calculated value of the Kolmogorov entropy λ^+, characterizing the mean divergence of trajectories from the invariant set, turned out to be 0.0355. The influence of relativistic effects on the picture of stochastization for $|\dot{x}|\simeq 1$ was also analyzed. Instead of (5.20) the equation

$$\frac{\mathrm{d}^2x}{\mathrm{d}t^2}=E\frac{(1-\dot{x}^2)^4}{1+5\dot{x}^2}\sin x\sin t \tag{5.22}$$

was examined. Relativistic effects turned out not to hinder chaotization of the soliton motion. Note that the influence of the radiated field on soliton stochastization remains as yet unstudied, though Ref. [5.4] contains first estimates of taking radiation field effects into account.

Actually, the adiabatic equations are valid for $t<\varepsilon^{-1}$. However, the stochasticity developed is realized on times of the same order. Therefore, it is desirable to take account of terms of the next order in ε, as this allows extension of the region of applicability up to times $t \sim \varepsilon^{-2}$, and so on. As is shown in Ref. [5.6], taking account of order ε^2 leads to the following closed equation for the soliton center coordinate x_s:

$$\frac{d^2 x_s}{dt^2} = -2\left[1 + \tfrac{8}{3}c^2 \frac{1 + 5x_s^2}{(1 + \dot{x}_s^2)^4}\right]^{-1} \left\{\tilde{n}_x - \frac{c}{(1 - \dot{x}_s^2)^4}\right.$$
$$\left. \times \left[(1 + 6\dot{x}_s^2 + \dot{x}_s^4)\dddot{x}_s + \zeta \dot{x}_s(\ddot{x}_s)^2 \frac{5 + 10\dot{x}_s^2 + \dot{x}_s^4}{(1 - \dot{x}^2)^2}\right]\right\}. \tag{5.23}$$

The terms like $\dddot{x}$ define the radiation field effects on the soliton motion and are of dissipative character. The radiation field effect on the soliton is equal to the energy (with the opposite sign), carried away by this field. It can be concluded that with the second-order corrections taken into account, the character of dynamic stochasticity in the soliton motion can be changed, to lead to dynamics of the type represented by the strange attractor.

5.3 DYNAMICAL CHAOS OF BREATHERS IN AN EXTERNAL FIELD

Here we consider the influence of a periodic external force, and of dissipation, on SG breather dynamics [5.8], [5.10].

The wave equation takes the form of a perturbed sine–Gordon equation (notation is the same as in Ref. [5.9]),

$$\varphi_{tt} - \varphi_{xx} + \sin\varphi = f(t) - \alpha\varphi_t, \tag{5.24}$$

where $\varphi = 2\pi\Phi/\Phi_0$ is the magnetic flux normalized by the value of its quantum $\Phi_0 = \pi c/(2e)$; x, t are dimensionless coordinate and time in standard variables; f is current density, and α is a coefficient proportional to the junction electrical conductivity. For $f = \alpha = 0$, a bion or breather is described by the expression

$$\varphi = -4\tan^{-1}\left\{\frac{\sinh[r(t)/2]}{\cosh x}\right\}, \tag{5.25}$$

where $r(t)$ is the soliton–antisoliton distance. It is assumed that $r(t) \gg 1$. At such distances, the bion is a superposition of soliton and antisoliton, together forming a coupled system with exponential attractive potential.

Let us formulate the basic equations to be applied in the following. Following a quasi-particle method [5.9], we come to the equation for the breather oscillations as

$$\frac{d^2r}{dt^2} = -8e^{-|r|} + (\pi/2)f(t) - \alpha\frac{dr}{dt}. \tag{5.26}$$

For $\alpha = f = 0$ the oscillator energy is

$$E = p^2/2 + U(r), \qquad U(r) = -8e^{-|r|}, \quad p \equiv dr/dt.$$

It is seen that for $E < 0$ the soliton–antisoliton pair is in the bound state, while for $E > 0$ it is in the free state. Breather oscillations ($E < 0$) are described by the formula

$$r(t) = \ln\{(8/|E|)\cos^2(\sqrt{(|E|/2)}t)\}. \tag{5.27}$$

We proceed first of all with the action-angle variables (I, θ) to describe the breather's motion in the region of $E < 0$. Oscillation frequency as follows from (5.27) is

$$\omega(H_0) = \frac{\sqrt{|H_0|}}{8\cos^{-1}\sqrt{(|H_0|/8)}}, \qquad E = H_0. \tag{5.28}$$

Hence the action variable is derived in the form

$$I = 1/(2\pi)\oint p\,dr = I_s\left\{\sqrt{(1-|H|/8)} - \sqrt{(|H|/8)}\tan^{-1}\frac{2\sqrt{[2\sqrt{(1-|H|/8)}]}}{\sqrt{|H|}}\right\}, \tag{5.29}$$

where I_s is the action on the separatrix. Starting from the fact that chaos occurs in the vicinity of the separatrix, we will consider the region where

$$\Delta I = |I - I_s| \ll I_s, \quad I_s = 8/\pi, \quad H(I_s) = 0,$$
$$H(I) \simeq -2^7\pi^2(1 - I/I_s)^2.$$

It is easy to see that

$$\omega(I) \simeq 2^8\pi^2(1 - I/I_s), \qquad \theta = \omega(I)t + \theta_0. \tag{5.30}$$

The equations of motion read as follows:

$$\frac{dI}{dt} = -f_0\frac{\partial V}{\partial \theta}, \qquad \frac{d\theta}{dt} = \omega(I) + f\frac{\partial V}{\partial I}.$$

The total Hamiltonian in the variables (I, θ) is

$$H(I,\theta) = H_0(I,\theta) + f_0 V(I,\theta) = H_0(I,\theta) + f_0 \sum_{n=-\infty}^{\infty} V_m \mathrm{e}^{\mathrm{i}m\theta + \mathrm{i}n\Omega t}. \quad (5.31)$$

Here the perturbed Hamiltonian is

$$V(I,\theta) = (\pi/2)\cos(\Omega t) r(I,\theta).$$

For the following analysis we need to define the form of the coefficients V_{ms} in the vicinity of the separatrix $(I \to I_{\mathrm{s}})$. The expression for r is expanded in series to estimate the behavior of expansion coefficients. Then we obtain the following formula for V_{m}:

$$V_{\mathrm{ms}} \simeq \frac{(-1)^m}{m}, \qquad m \gg 1. \quad (5.32)$$

As is seen from (5.31), upon the action of the periodic force on the breather, the oscillations in the latter exhibit nonlinear resonances. The condition for such resonances to occur is

$$2m\omega(I) = \Omega. \quad (5.33)$$

The resonance separation in the vicinity of the separatrix is

$$\delta\omega = |\omega_{m+1} - \omega_m| = \frac{\Omega}{2m^2} \simeq \frac{2\omega^2}{\Omega}. \quad (5.34)$$

With the aid of (5.32) we obtain the width of the nonlinear resonance as

$$\Delta\omega = \max|\omega(I) - \omega(I_0)| = \left|4 f_0 V_{\mathrm{m}} \frac{\mathrm{d}\omega}{\mathrm{d}I}\right|^{1/2} \simeq \frac{16}{\pi I_{\mathrm{s}}}(f_0\omega/\Omega)^{1/2}. \quad (5.35)$$

Overlapping of nonlinear resonances leads to stochastic oscillations of the coupled soliton–antisoliton system. The limiting stochasticity can be evaluated using Chirikov's criterion for resonance overlapping. Making use of (5.34) and (5.35), we obtain the condition

$$K = \left(\frac{\delta\omega}{\Delta\omega}\right)^2 > 1, \quad \frac{2^7}{\pi^2 I_{\mathrm{s}}^2} \frac{f_0\Omega}{\omega^3(I)} > 1. \quad (5.36)$$

This shows that for any f_0 and Ω the parameter K grows with I approaching the separatrix $(\omega \to 0, I \to I_{\mathrm{s}})$. Thus, for $\omega = \bar{\omega}$ a stochastic layer is formed. Breather oscillations are random in it, with

$$\omega \leqslant (f_0\Omega)^{1/3}\frac{2^{7/3}}{(\pi^2 I_s)^{1/3}}. \tag{5.37}$$

With damping taken into account we can estimate the limiting stochasticity, resorting to qualitative arguments from Ref. [5.5]. The condition for chaos to occur is

$$K_\mu \geqslant 1, \qquad \mu = \frac{1-\exp(-\alpha)}{\alpha}. \tag{5.38}$$

For $\alpha \gg 1$, $K/\alpha \gg 1$. Randomness in this case is described by behavior of strange attractor type. Let us examine the behavior of this system in the phase plane (Figure 5.2). For $E < 0$ there exists the bound state. The region of $E > 0$ describes the infinite motion corresponding to breather decay into a free soliton–antisoliton pair. Under periodic perturbation about the separatrix, a stochastic layer of width $\sim (f_0\Omega)^{1/3}$ is formed. Breather oscillations are random within this layer.

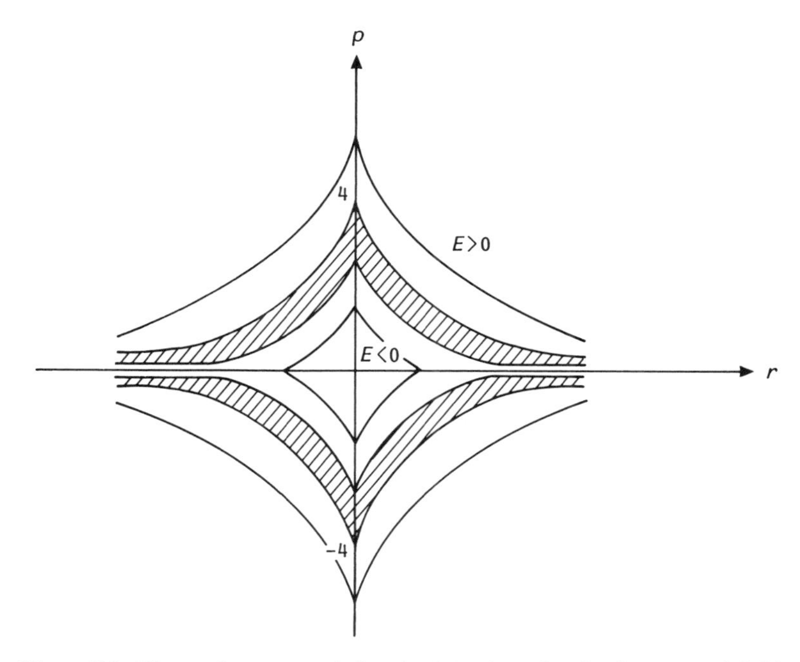

Figure 5.2. Phase-plane portrait for the SGE breather in the external field

Now we will carry out the estimation needed for the case of a long Josephson junction in a periodic field. Consider the case of $f_0 = 0.01$, $\Omega = 1$. From (5.37) we derive the time of breather traverse over the junction, subsequent to which stochastic breather oscillations start. Thus, $t_c = 2\pi/\bar{\omega} \simeq 50\text{--}100$, i.e. stochasticity develops rather quickly. The soliton and antisoliton chaotic motion ought to give rise to an electromagnetic noise field radiated by the junction. With increasing f_0 one first should expect the generation of subharmonics, and only for $f_0 > f_{\text{crit}}$ the generation of a broadband spectrum.

The authors of Ref. [5.10] were the first to obtain (5.36), by an averaging method. Recently (see [5.11]), this estimate was improved to be $|\zeta|^3 \leqslant \pi^{2/2}\omega(\log \omega^{-1})|\varepsilon|$. Note that the quasi-particle method applied has its own advantages as it allows the description of stochastic dynamics of the soliton bound states in *nonintegrable* systems.

Further, we will now analyze the process of stochastic breather decay in an external field [5.11]. For this we consider the evolution of a weakly bound breather in an external periodic field having the form $f(t) = \varepsilon \sin \omega t$. Applying IST-based perturbation theory we obtain the equation for the parameter ζ^2 as

$$\frac{d\zeta^2}{dt} = \varepsilon \sin \omega t \sqrt{(1 - \zeta^2 T^2)}\, T[(1 + T^2)^{-1} + T^{-1}(1 + T^2)^{-3/2} \sinh^{-1} T]. \tag{5.39}$$

where $\zeta \equiv \pi/2 - \mu$ and μ is the breather amplitude. The main contribution to the change of ζ^2 is given by the region in which kink and antikink have significant overlap. Overlapping takes place at times which are a small fraction of the total period of breather oscillations, $\tau = 2\pi/\zeta$. The times at which the soliton and antisoliton largely overlap are given by

$$\zeta^2 T^2 \ll 1, \qquad dT/dt \simeq 1.$$

Then equation (5.39) takes the form

$$\frac{d\zeta^2}{dT} = \varepsilon[\sin(\omega T + \delta) T[(1 + T^2)^{-1} + T^{-1}(1 + T^2)^{-3/2} \sinh^{-1} T], \tag{5.40}$$

where δ is the phase difference between the intrinsic breather oscillations and the external field at the moment of maximal overlap. The jump ζ^2 arising from a single overlap is found to have the form

$$\Delta\zeta^2 \simeq \int_{-\infty}^{\infty} \frac{d\zeta^2}{dT}\, dT \simeq \varepsilon \cos \delta g(\omega), \tag{5.41}$$

where

$$g(\omega) = 2\int_{-\infty}^{\infty} \mathrm{d}T(\sin\omega T)(1+T^2)^{-1}[T+(1+T^2)^{-1/2}\sinh^{-1}T].$$

In particular, for $\omega \ll 1$, we have

$$g(\omega) = \pi + \frac{\pi^2}{2}\omega^2 \operatorname{sgn}\omega + O(\omega^3),$$

and for $\omega \gg 1$

$$g(\omega) \sim \mathrm{e}^{-\omega}.$$

In contrast to ζ^2, which grows under the action of kicks, the phase difference increases uniformly, so that the change of δ within the half-period between two kicks is described as follows:

$$\Delta(\delta) \simeq \pi(\omega/\zeta - 1). \tag{5.42}$$

Then instead of differential equations (5.39) and (5.40) we have a system of discrete mappings:

$$\begin{aligned} \zeta_{n+1}^2 &= \zeta_n^2 + \varepsilon g(\omega)\cos\delta_n, \\ \delta_{n+1} &= \delta_n + \pi(\omega/\zeta_{n+1} - 1). \end{aligned} \tag{5.43}$$

Here n is the number of overlappings. This mapping is well known: it appears in some branches of plasma physics, etc. Using standard results, we find that

$$|\zeta^3| \leqslant \omega g(\omega)|\varepsilon|.$$

5.4 STOCHASTIC DYNAMICS OF A BREATHER INTERACTING WITH A LOCAL INHOMOGENEITY

We examine here oscillations of a weakly bound breather trapped at a local inhomogeneity [5.11]. The perturbation in this case is

$$R(\varphi) = \delta(x)\sin\varphi. \tag{5.44}$$

The total breather Hamiltonian in the presence of the perturbation (5.44) has, in the first approximation, the form

$$\begin{aligned} H = 16\sin\mu(1+v^2/2) - 8\varepsilon\tan^{-2}(\mu)\sin^2\Psi\cosh^2\varphi \\ (\sin^2\Psi + \tan^{-2}\mu\cosh^2\varphi)^{-1}, \end{aligned} \tag{5.45}$$

where

$$\varphi = \zeta \sin \mu, \qquad (v^2 \ll 1).$$

It will be assumed in the following that the characteristic perturbation time $\tau\rho \simeq |\varepsilon|^{-1/2}$ is much greater than the period of breather oscillations $\tau\beta \sim 2\pi/\cosh \mu$. Then the total Hamiltonian (5.45) can be averaged over fast oscillations, to result in

$$\begin{aligned}\langle H\rangle + \langle H_1\rangle &= 16 \sin \mu(1 + v^2/2) - 4 \tan^{-1} \mu \cosh \varphi(1 + \tan^{-2} \mu \cosh^2 \varphi)^{-3/2} \\ &\quad - 8\varepsilon \tan^{-1} \mu \cosh \varphi(1 + \tan^{-2} \mu \cosh^2 \varphi)^{-3/2} \\ &\quad + [1 - 4 \tan^{-1} \mu \cosh \varphi(1 + \tan^{-2} \mu \cosh \varphi)^{3/2} \\ &\quad + 2 \tan^{-2} \mu \cosh^2 \varphi(3 + 2 \tan^{-2} \cosh^2 \varphi)] \tan^{-2}[2(\cos \mu + \Psi_0)].\end{aligned} \tag{5.46}$$

The first two terms are the intrinsic averaged $\langle H\rangle$, and the third term H_1 is the first oscillating correction (note that $\mu \ll 1$). The term H_1 describes dynamics of the effective particle of mass $16 \sin \mu$ moving in the potential

$$U = -8\varepsilon \tan^{-1} \mu \cosh(\sin \mu \cdot \zeta)[1 + \tan^{-2} \mu \cosh^2(\sin \mu \cdot \zeta)]^{-3/2}. \tag{5.47}$$

Consider now $\xi \equiv \pi/2 - \mu \ll \sqrt{\varepsilon}$. in this case we can apply an unaveraged

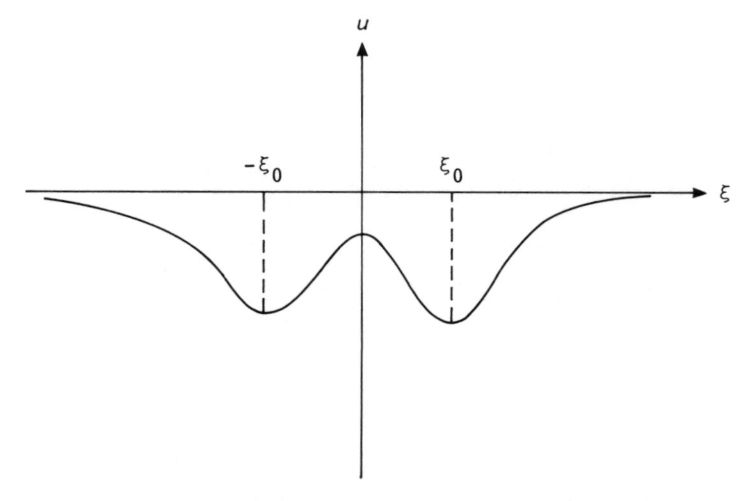

Figure 5.3. Effective potential for the breather interacting with a local inhomogeneity [5.11]

Hamiltonian that with the aid of the variables v, φ, ζ and $T = \zeta^{-1}\sin\Psi$, can be written in the form

$$H \approx 16(1 - \zeta^2/2 + v^2/2) - 8\varepsilon T(T^2 + \cosh^2\varphi)^{-2}\cosh\varphi. \qquad (5.48)$$

For $\varepsilon > 0$ the particle is seen to oscillate inside either of two potential wells (Figure 5.3):

$$U = -4\varepsilon e^z(4 + e^{2z})^{-3/2}, \qquad z \equiv |\zeta| - \log\zeta^{-1}$$

spaced by the distance

$$L = 2\log\zeta - 1.$$

The application of the criterion for resonance overlapping shows that the condition for stochasticity to occur is $\zeta \leqslant \sqrt{\varepsilon}$. In the case of $\varepsilon < 0$ this condition is

$$\zeta \leqslant \sqrt{|\varepsilon|}.$$

5.5 DYNAMICAL CHAOS OF A BREATHER IN PARAMETRICALLY DRIVEN SG SYSTEMS

Let us consider now a parametrically perturbed SG system. The corresponding equation is

$$\varphi_{tt} - \varphi_{xx} + (1 + \varepsilon_0\cos\Omega t)\sin\varphi = 0. \qquad (5.49)$$

Our goal is to study the dynamics of an initial state taken as a single breather [5.12], [5.13]. For $\varepsilon = 0$ the breather solution has the form

$$\varphi_B(z) = -4\tan^{-1}\left(\frac{\nu}{\eta}\frac{\cos\varphi}{\cosh z'}\right),$$

$$\varphi = \theta(z) - (\eta/|\nu|)vz, z = \frac{\nu}{|\zeta|}\frac{x - x_0}{\sqrt{(1 - v^2)}},$$

$$v = \frac{1 - 4|\zeta|^2}{1 + 4|\zeta|^2}, \qquad \zeta = \eta + i\nu, \quad \eta > 0, \quad \nu > 0. \qquad (5.50)$$

Here v is the breather velocity, $d\theta/dt$ is the frequency of the intrinsic breather oscillations. Using perturbation theory for breathers (A.54–57) [5.13], we obtain the following equation for the slowly time-varying breather parameters $\gamma(t) = \tan^{-1} v/\eta$ and $\theta(t)$:

$$\frac{\mathrm{d}\gamma}{\mathrm{d}t} = \frac{8\varepsilon(t)\sin\theta}{\cos\gamma(1+A^2)^2}\left\{3A + \frac{1-2A^2}{\sqrt{(1+A^2)}}\sinh^{-1} A\right\}, \tag{5.51}$$

$$\frac{\mathrm{d}\theta}{\mathrm{d}t} = \frac{8\varepsilon(t)\sin^2\theta}{\cos^3\gamma(1+A^2)^2}\left\{2 + \sin^2\gamma\sin^2\theta + \frac{\mathrm{A}\sinh^{-1} A}{\sqrt{(1+A^2)}}(\tan^2\gamma\cos^2\theta - 2)\right\}, \tag{5.52}$$

$$A = \tan\gamma\cos\theta.$$

Note that the variables are in reality action-angle variables. The equation for the breather velocity is not written out, since it is unchanged under the perturbation $\cos(\Omega t)\sin\varphi$.

Examine now the behavior of solutions for the set (5.51) and (5.52). This set is of Hamiltonian type. Its Hamiltonian has the form

$$H(\gamma, \theta) = H_0 + V_{\mathrm{inf}},$$

$$H_0 = \sin\gamma,$$

$$V_{\mathrm{int}} = \frac{8\varepsilon(t)}{\sin\gamma(1+A^2)}\left\{A + \frac{\sinh^{-1} A}{\sqrt{(1+A^2)}}\right\}. \tag{5.53}$$

The frequency of intrinsic breather oscillations in variables (γ, θ) is $\omega(\gamma) = \mathrm{d}H_0/\mathrm{d}\gamma = \cos\gamma$. The equations of motion (5.51) and (5.52) have the forms

$$\frac{\mathrm{d}\gamma}{\mathrm{d}t} = -\varepsilon\frac{\partial V}{\mathrm{d}t}, \qquad \frac{\mathrm{d}\theta}{\mathrm{d}t} = \omega(\gamma) + \varepsilon\frac{\partial V}{\partial\gamma}.$$

Following the procedure of Ref. [5.2] we will analyze the breather oscillations. To begin with, the perturbation Hamiltonian is expanded in the Fourier series

$$V(\gamma, \theta, t) = \tfrac{1}{2}\sum_{m=-\infty}^{\infty} V_m(\gamma) l^{im\theta + \mathrm{i}\Omega t}. \tag{5.54}$$

This shows that the strongest perturbation effect on the breather arises when

$$m\omega(\gamma) + \Omega = \theta, \tag{5.55}$$

i.e. for nonlinear resonance. The domain of nonlinear resonance localization is

$$\Delta\gamma = \max|\gamma - \gamma_0| = 4\left|\varepsilon V_m \left(\frac{\mathrm{d}\omega}{\mathrm{d}\gamma}\right)^{-1}\right|^{1/2},$$

$$\Delta\omega = \max|\omega - \omega_0| = \Delta\gamma\frac{\mathrm{d}\omega}{\mathrm{d}\gamma} = 4\left|\varepsilon V_{\mathrm{m}}\frac{\mathrm{d}\omega}{\mathrm{d}\gamma}\right|^{1/2}. \tag{5.56}$$

Let us determine the width of the nonlinear resonance for a small ($\gamma \to 0$) and a breather of large ($\gamma \to \gamma_s = \pi/2$) amptitude. Here $\gamma = \gamma_s$ is the value of the action on the separatrix, separating the breather states from the free kink–antikink system.

(a) A small amplitude breather ($\gamma \to 0$). We have resonance for $m = 1$, $\omega(\gamma_1) = \Omega/2, \gamma_1 = \cos^{-1}(\Omega/2)$. V_m is estimated for ($\gamma \to 0$) to have the form

$$V_m \simeq \frac{\tan^2\gamma}{8}(-1)^{m/2}. \tag{5.57}$$

Thence we obtain

$$\frac{\Delta\gamma}{\gamma_s} = \frac{(2\sqrt{2})\varepsilon_0^{1/2}}{\pi}\frac{\sin^{1/2}\gamma}{\cos\gamma},$$

$$\Delta\omega = (\sqrt{2})\varepsilon_0^{1/2}\frac{\sin^{3/2}\gamma}{\cos\gamma}.$$

(b) A large amplitude breather. In the vicinity of the separatrix ($\gamma \to \pi/2$), we have

$$H_0(\gamma_s) = 1, \qquad \omega(\gamma_s) = 0.$$

For $m \gg 1$, $m \ll \tan\gamma$, we obtain the estimate

$$V_{\mathrm{m}} \simeq \frac{(-1)^n}{a}. \tag{5.58}$$

Then from (5.56) and (5.58) we derive

$$\frac{\Delta\gamma}{\gamma_s} = \frac{8\varepsilon_0^{1/2}\cos^{1/2}\gamma}{\pi\sin\gamma},$$

$$\Delta\omega \simeq 4\varepsilon_0^{1/2}\cos^{1/2}\gamma \simeq 4\sqrt{(\varepsilon_0\omega)}. \tag{5.59}$$

The results obtained allow us to find the conditions for random breather oscillations to occur under the action of fluctuating fields. Note that the results of Ref. [5.2] indicate the occurrence in the vicinity of the separatrix of a stochastic layer in which breather oscillations are random, the stochastic layer occurring under periodic perturbations of random character. The region of stochasticity can be defined starting from the following considerations (Chirikov's criterion). One can easily see that in vicinity of the separatrix the resonance spacing decreases with increasing m. Indeed,

$$\delta\omega = \max|\omega_{m+1} - \omega_m| = \frac{\Omega}{m^2} \simeq \frac{2\cos 2\gamma}{\Omega}. \tag{5.60}$$

It can be assumed that when the width of nonlinear resonances is so large that the distance between them leads to resonance overlapping, the motion in this region will be of stochastic character. Then the stochasticity criterion (Chirikov's criterion) has the form

$$K = \left(\frac{\Delta\omega}{\delta\omega}\right)^2 \geqslant 1. \tag{5.61}$$

Using (5.60) and (5.61) we obtain

$$K = \frac{4\varepsilon_0\Omega^2}{\cos^3\gamma} > 1. \tag{5.62}$$

For any ε_0 and Ω with $\gamma \to \gamma_s$, the parameter k is seen to increase (as $\omega(\gamma) = \cos\gamma \to 0$), and there is always a region of values of γ available wherein the condition (5.62) is met. The boundary of the stochastic layer is given by $\bar{\gamma}$, where

$$\bar{\gamma} = \cos^{-2}(4\varepsilon_0\Omega^2)^{1/3}, \tag{5.63}$$

and its width near the separatrix is

$$|\gamma_s - \bar{\gamma}| = (4\varepsilon_0\Omega^2)^{1/3}, \qquad (\varepsilon_0 \ll 1, \Omega \leqslant 1). \tag{5.64}$$

Should the initial breather parameters be such as to lie in the region of parameters satisfying (5.5.16) (the dashed area in Figure 5.4), its oscillations become random and its dynamics of diffusive character. As a result, the breather disintegrates stochastically into the free domain walls. The time of breather disintegration is of the order

$$\tau = \frac{2\pi}{\omega} \simeq \frac{2\pi}{(4\varepsilon_0\Omega^2)^{1/3}},$$

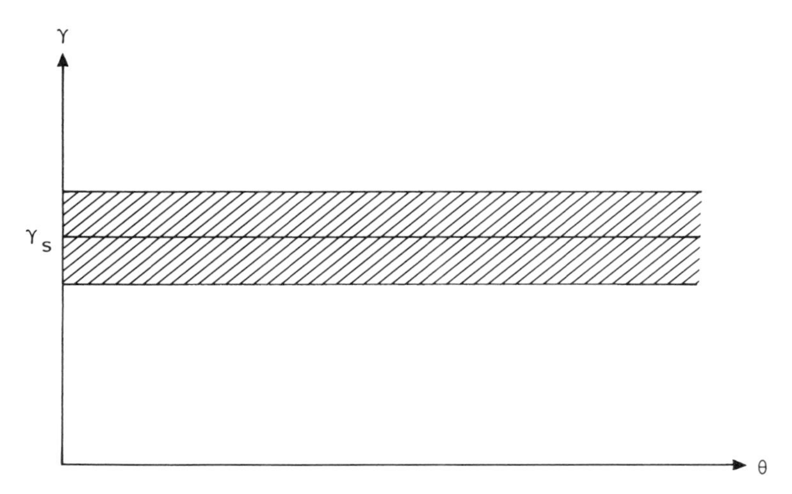

Figure 5.4. Phase portrait for the SGE parametrically driven breather

and for $\varepsilon_0 \simeq 10^{-2}$, $\tau \simeq 10t_\Omega$. Here the estimates are for the frequently used quasi-one-dimensional ferromagnetic $CsNiF_3$ placed in *a* the high-frequency field [5.14]. Under the experimental conditions we have $I = 11.2\,K$, $A \simeq 2.3\,K$, $a = 5 \times 10^{-7}$ cm, $H_0^x = 5$ kGs. From (5.64) we infer that magnetization chaos occurs for

$$h_0 = 50\,G, \quad \Omega = 120\,MHz, \quad \gamma/\eta \simeq 10.$$

In conclusion we note that a more comprehensive review of dynamical chaos of solitons and nonlinear periodic waves is contained in Ref. [5.15].

Chaos of SG and NLS solitons interacting with an oscillatory boundary or impurity has been investigated in Refs. [5.16], [5.17]. It is solitonic analog of Ulam model for Fermi acceleration [5.5]. It is necessary to note, that other separatrix (or homoclinic) modes of the nonlinear evolution equation are important for chaos [5.18]. Some numerical experiments reported in Refs. [5.19], [5.20] support this point of view. Attempts to explain the observed phenomena have been carried out [5.21] using the Melnikov function method [5.18]. For other developments see Refs. [5.22], [5.23].

6

CONCLUSION

In this book we have tried to report the main aspects of the theory of nonlinear waves in inhomogeneous media. Progress in the theoretical development is created by the possibility to separate localized wave packets of the soliton and breather types and nonlocalized modes of periodic wave type. The description of nonlinear wave evolution during propagation is reduced to a finite-dimensional description. Various asymptotic methods, in particular perturbation theory based on the inverse scattering transform, appear to be very efficient. The analysis has shown that solitons in inhomogeneous media behave as particles. Moreover there exist, together with pure Newtonian dynamics, the novel processes of wave emission by solitons, of soliton decay, and of new soliton generation during solitan passage through inhomogeneities.

Autowave processes in inhomogeneous nonlinear media, in particular prediction of soliton and breather existence in active media, are also of great interest.

The stochastic dynamics of solitons in inhomogeneous media is also possible, together with regular soliton dynamics. The stochastic dynamics can be induced by the random field influence and by system inhomogeneity. There exists another type of random dynamics of solitons and other nonlinear waves, caused by a strong instability of a system for certain system and wave parameters (so-called dynamical chaos). The investigation of these problems has just been started, and one can expect rapid progress. One can hope that the solution of these problems will be helpful in the investigation of Langmuir turbulence, optical chaos in nonlinear waveguides, etc.

It is necessary to note that at present there are various problems that require further investigations. First of all, there is the investigation of phenomena dealing with wave emission by solitons, and strong soliton coupling with a field of emission. This requires a development of analytical methods, being outside the scope of perturbation theory.

The next problem is a description of the propagation of soliton bound states and nonlinear periodic waves in inhomogeneous media, and also analysis of soliton interaction in inhomogeneous and active media. As to the last problem, we face it while describing N-soliton complex transmission in long optical communication cable systems with periodic amplification.

We believe that the investigation of the theory of solitons in inhomogeneous media will develop rapidly in the near future.

Appendix 1

THE NONLINEAR SCHRODINGER EQUATION (NLS)

In this Appendix we will give reference material on the NLSE theory required for reading this book. The NLSE has the form

$$iq_t + q_{xx} + 2|q|^2 q = 0. \tag{A.1}$$

Let us suppose

$$\int_{-\infty}^{\infty} |q_0(x, 0)| \, dx < \infty. \tag{A.2}$$

The solution of the initial-value problem for rapidly vanishing initial conditions (A.1) is reduced to the investigation of a spectrum for a set of two differential first-order equations (Zakharov–Shabat system)

$$\begin{aligned} \psi_x^{(1)} &= i\lambda\psi^{(1)} + iq_0(x)\psi^{(2)}, \\ \psi_x^{(2)} &= -i\lambda\psi^{(2)} + iq_0^*(x)\psi^{(1)}. \end{aligned} \tag{A.3}$$

Discrete values of λ_n are representative of solitons and N-soliton states, formed during the evolution of a rapidly vanishing initial condition; continuous values of λ are responsible for a continuous wave excited in the system. We will specify the most characteristic nonlinear modes used in this book.

(a) *One-soliton state.* A relevant spectral parameter is $\lambda_1 = \mu + i\eta$. The one-soliton solution has the form

$$q_s(x, t) = 2\eta \operatorname{sech} 2\eta(x - x_0 + 4\mu t) \exp\{-2i\mu x - 4i(\mu^2 - \eta^2)t - i\delta_0\} \tag{A.4}$$

Here η is the soliton amplitude, $v = -4\mu$ its velocity.

(b) *Two-soliton state.* It is written as follows:

$$q(t,x) = 4\exp(-it/2)[\cosh(4x) + 4\cosh(2x) + 3\cos(4t)]^{-1} \times [\cosh(3x) + 3\exp(-4it)\cosh(x)]. \qquad (A.5)$$

The modulus of this solution is periodic in t, with the period $= \pi/2$.

(c) *The case of large smooth initial potential.* Let the initial potential satisfy the condition $q_x/q \ll \lambda$. Then the values of η_n are defined by the Bohr quantization rule

$$\int_{-\infty}^{\infty} \sqrt{(|q_0|^2 + \eta_n^2)}\, dx = 2\pi(n + \tfrac{1}{2}). \qquad (A.6)$$

The total number of levels, which coincides with the number of solitons is

$$N = 1/\pi \int_{-\infty}^{\infty} |q_0(x)|\, dx. \qquad (A.7)$$

The condition for absence of solitons has the form

$$\int_{-\infty}^{\infty} |q_0(x)|\, dx < \ln(2 + \sqrt{3}) \qquad (A.8)$$

There exist an infinitely many of integral invariants of NLSE. The first three are the so-called 'number of quanta' N, the field momentum P, and the total Hamiltonian H,

$$N = \int_{-\infty}^{\infty} |q|^2\, dx, \qquad (A.9)$$

$$P = i/2 \int_{-\infty}^{\infty} (q^* q_x - q_x^* q)\, dx, \qquad (A.10)$$

$$H = \int_{-\infty}^{\infty} (|q_x|^2 - |q|^4)\, dx. \qquad (A.11)$$

For the case of a single soliton, substitution of (A.4) into (A.9–11) yields

$$N = 4\eta, \quad P = 8\mu\eta, \quad H = 32\eta((1/3)\eta^2 - \mu^2). \qquad (A.12)$$

The Jost functions for the single-soliton initial state have the form

$$\Psi = \begin{pmatrix} \Psi^{(1)} \\ \Psi^{(2)} \end{pmatrix} = \frac{e^{i\lambda x}}{\lambda - \mu + i\eta} \begin{pmatrix} \lambda - \mu + i\eta \tanh 2\eta(x - x_0) \\ \eta\, \mathrm{sech}\, 2\eta(x - x_0) e^{-2i\mu x - i\delta_0} \end{pmatrix}, \qquad (A.13)$$

$$\phi(x,\lambda) = a(\lambda)\tilde{\psi}(x,\lambda), \tag{A.14}$$

$$a(\lambda) = \frac{\lambda - \mu - i\eta}{\lambda - \mu + i\eta},$$

$$\tilde{\Psi} = \begin{pmatrix} -\Psi^{(2)*} \\ \Psi^{(1)*} \end{pmatrix}. \tag{A.15}$$

The 'action-angle' variables are of the following form:

$$n(\lambda) = \frac{1}{\pi}\ln\frac{1}{|a(\lambda)|^2} \tag{A.16}$$

$$\phi(\lambda) = \arg b(\lambda) \tag{A.17}$$

$$N_k = 2\lambda_k, \quad \Phi_k = \ln(1/b_k), \qquad k = 1, 2, \ldots, N.$$

Introducing new variables

$$\bar{N}_k = i(N_k^* - N_k), \qquad \mu_k = (\Phi_k^* - \Phi_k)/2i, \tag{A.18}$$

we define motion integrals in these variables as

$$N = \sum_k \bar{N}_k + \int_{-\infty}^{\infty} n(\lambda)\,d\lambda,$$

$$P = i\int_{-\infty}^{\infty} qq_x^*\,dx = \sum_k \frac{v_k \bar{N}_k}{2} - 2\int_{-\infty}^{\infty} \lambda n(\lambda)\,d\lambda \tag{A.19}$$

$$E = H = \sum_k\left(-\frac{\bar{N}_k^3}{12} + \frac{\bar{N}_k v_k^2}{4}\right) + 4\int_{-\infty}^{\infty} \lambda^2 n(\lambda)\,d\lambda$$

Here $v_k = -4\,\mathrm{Re}\,\lambda_k$ is the velocity of the kth soliton, $N_k/2$ is the soliton mass, $\bar{N}_k v_k^2/4$ is the kinetic soliton energy, and $\bar{N}_k^3/12$ is the rest energy. The magnitude $(n(\lambda/2))\,d\lambda$ is interpreted as the mass of particles moving with velocities in the range $(-4\lambda, -4(\lambda + d\lambda)$.

In nonlinear optics, condensed matter, for normal dispersion $d^2\omega/dk^2 > 0$. The NLS is then of the other sign of nonlinearity,

$$iq_t + q_{xx} - 2|q|^2 q = 0, \tag{A.20}$$

It has a 'dark' soliton solution

$$q_s(x,t) = 2\eta\tanh 2\eta(x - x_0 + 4\mu t)\exp\{-2i\mu x - 4i(\mu^2 + 2\eta^2)t - i\delta_0\}. \tag{A.21}$$

Let us report the formulas describing the evolution of the NLS solitons under weak perturbations

$$iq_t + q_{xx} + 2|q|^2 q = i\varepsilon R(q), \qquad \varepsilon \ll 1. \tag{A.22}$$

The one-soliton solution is chosen in the form

$$q_s(x,t) = 2\eta(t)\,\mathrm{sech}[2\eta(x-\zeta(t))]\exp\{iz\mu(t)/\eta(t) + i\delta(t)\}, \tag{A.23}$$

where $z = 2\eta(x-\zeta)$. The application of perturbation theory to solitons, based on the [A1]; [A2], allows the derivation in the adiabatic approximation of the following equations for the soliton parameters:

$$\frac{d\mu}{dt} = \tfrac{1}{2}\varepsilon \,\mathrm{Im}\left\{\int_{-\infty}^{\infty} dz\, R\, e^{-i\theta}\,\mathrm{sech}\, z \tanh z\right\}, \tag{A.24}$$

$$\frac{d\eta}{dt} = \tfrac{1}{2}\varepsilon\, \mathrm{Re}\left\{\int_{-\infty}^{\infty} dz\, R\, e^{-i\theta}\,\mathrm{sech}\, z\right\}, \tag{A.25}$$

$$\frac{d\zeta}{dt} = 4\mu + \frac{\varepsilon}{4\eta^2}\mathrm{Re}\left\{\int_{-\infty}^{\infty} dz\, z\,\mathrm{sech}\, z\, R\, e^{-i\theta}\right\}, \tag{A.26}$$

$$\frac{d\delta}{dt} = 2\mu\frac{d\zeta}{dt} - 4(\mu^2 - \eta^2) + \frac{\varepsilon}{2\eta}\mathrm{Im}\left\{\int_{-\infty}^{\infty} dz\,\mathrm{sech}\, z(1 - z\tanh z)\, R\, e^{-i\theta}\right\},$$
$$\theta = (\mu/\eta)z + \delta. \tag{A.27}$$

Let us also write out the equations of perturbation theory for the evolution of nonlinear wave packets in a soliton-free sector [A3]. The perturbed evolution equations for the 'action-angle' variables $n(\lambda)$ and $\varphi(\lambda)$ are

$$dn/dt = 2/\pi \int_{-\infty}^{\infty} dx\,\mathrm{Im}\{b(\lambda)/a(\lambda)[\varepsilon(\psi_2^*(x,\lambda))^2 R(x) - \varepsilon^*(\psi_1^*(x,\lambda))^2 R^*(x)]\}, \tag{A.28}$$

$$d\varphi/dt = 4\lambda^2 + \mathrm{Re}\left\{\int_{-\infty}^{\infty} dx\,\{a(\lambda)/b(\lambda)[\varepsilon^*(\psi_2^*(x,\lambda))^2 R^*(x) - \varepsilon(\psi_1(x,\lambda))^2 R(x)] - 2\varepsilon\psi_1(x,\lambda)\psi_2^*(x,\lambda)R(x)\right\}. \tag{A.29}$$

For specific initial conditions we usually take

(a)

$$q_0(x)=\begin{cases} a_0, & 0\leqslant x\leqslant l_0,\\ 0, & x<0, x>l_0,\end{cases} \tag{A.30}$$

$$n(\lambda)=1/\pi\{1+[\sin^2(l_0\sqrt{(\lambda^2+a_0^2)})][\lambda^2/a_0^2+\cos^2(l_0\sqrt{(\lambda^2+a_0^2)})]^{-1}\}$$

(b)

$$q_0(x)=a_0\,\mathrm{sech}(x/l_0)\exp(\mathrm{i}gx),$$
$$n(\lambda)=-1/\pi\ln[1-\mathrm{sech}^2(\pi l_0(\lambda-g/2))\sin^2(\pi a_0 l_0)]. \tag{A.31}$$

We also report the expressions for variational derivatives: Jost coefficients and the spectral parameter λ. The equations for $a(\lambda,t), b(\lambda,t)$ can be obtained by using the relation

$$\frac{\mathrm{d}a}{\mathrm{d}t}=\int_{-\infty}^{\infty}\mathrm{d}x\left\{\frac{\delta a}{\delta q(x)}\frac{\partial q}{\partial t}+\frac{\delta a}{\delta q^*(x)}\frac{\partial q^*}{\partial t}\right\} \tag{A.32}$$

Substituting (A.22) in this equation we obtain

$$\frac{\mathrm{d}a}{\mathrm{d}t}=\left.\frac{\mathrm{d}a}{\mathrm{d}t}\right|_{\varepsilon=0}+\varepsilon\int_{-\infty}^{\infty}\mathrm{d}x\left\{\frac{\delta a}{\delta q(x)}R+\frac{\delta a}{\delta q^*(x)}R^*\right\}. \tag{A.33}$$

Taking into account the IST formulae for the evolution of the Jost coefficients $a(\lambda,t), b(\lambda,t)$ at $\varepsilon=0$,

$$a(\lambda,t)=a(\lambda,0),\quad b(\lambda,t)=b(\lambda,0)\exp(-4\mathrm{i}\lambda^2 t),$$

the expressions for variational derivatives (A.32) and expressions for single soliton Jost functions we have

$$\frac{\partial a}{\partial t}=\frac{\mathrm{i}\varepsilon}{(\lambda-\xi+\mathrm{i}\eta)^2}\mathrm{Re}\left\{\int_{-\infty}^{\infty}\mathrm{d}zR[u](\lambda-\xi-\mathrm{i}\eta\tanh z)\,\mathrm{sech}\,z\,\mathrm{e}^{\mathrm{i}\theta}\right\} \tag{A.34}$$

$$\frac{\partial b}{\partial t}=4\mathrm{i}\lambda^2 b-\frac{\mathrm{i}\varepsilon\exp(-2\mathrm{i}\xi\lambda)}{(\lambda-\xi)^2+\eta^2}\int_{-\infty}^{\infty}\frac{\mathrm{d}z}{2\eta}\{\eta^2\,\mathrm{sech}^2 z\exp(\mathrm{i}\theta-\mathrm{i}\lambda z/\eta)$$
$$R[u]-\exp(-\mathrm{i}\lambda z/\eta)(\lambda-\xi-\mathrm{i}\eta\tanh z)^2R^*[u]\} \tag{A.35}$$

For calculating the contribution of continuum spectrum excited by perturbation in the integrals of motion the Hamiltonian approach formulae (A.19) are useful. Small perturbation $n(\lambda)$ can be represented by

$$n|(\lambda) = (1/\pi)\ln(1/|a(\lambda)|^2) = (1/\pi)|b|^2$$

and from (A.19) it follows that

$$\frac{dN}{dt} = \sum_k \frac{dN_k}{dt} + \frac{2}{\pi}\int_{-\infty}^{\infty} d\lambda \mathrm{Re}\left[b\frac{db^*}{dt} \right]. \tag{A.36}$$

Using (A.35) and (A.36) we are able, for example, to find the inverse influence of radiative losses on soliton amplitude and other radiative effects.

The interaction of two solitons in the presence of an external perturbation inducing, for example, the media inhomogeneities, can be described by PT equations with perturbation in the following form:

$$\varepsilon R = \mathrm{i}(q_m^* q_n^2 + 2q_m q_n q_n^*) + \varepsilon R_1(q_m).$$

$$\frac{d\mu_n}{dt} = (-1)^n 16\eta^3 e^{-2\eta r}\cos(2\mu r + \psi) + \frac{\varepsilon}{2}\mathrm{Im}\int_{-\infty}^{\infty} \frac{\tanh z}{\cosh z} R_1[q_s]\, e^{-i\theta}\, dz$$

$$\frac{d\eta_n}{dt} = (-1)^n 16\eta^3 e^{-2\eta r}\sin(2\mu r + \psi) + \frac{\varepsilon}{2}\mathrm{Re}\int_{-\infty}^{\infty} \frac{R_1[q_s]}{\cosh z} e^{-i\theta}\, dz$$

$$\frac{d\xi_n}{dt} = 2\mu_n + 4\eta\, e^{-2\eta r}\sin(2\mu r + \psi) + \frac{\varepsilon}{4\eta^2}\mathrm{Re}\int_{-\infty}^{\infty} zR_1[q_s]\,\mathrm{sech}\, z\, e^{-i\theta}\, dz$$

$$\frac{d\delta_n}{dt} = 2(\eta^2 + \mu^2) + 8\mu\eta\, e^{-2\mu r}\sin(2\mu r + \psi)$$

$$+ \frac{\varepsilon}{2\eta}\mathrm{Im}\int_{-\infty}^{\infty} (1 - z\tanh z)\,\mathrm{sech}\, z R_1[q_s(z)]\, e^{-i\theta}\, dz$$

$$\eta = \tfrac{1}{2}(\eta_1 + \eta_2), \quad r = (\xi_1 - \xi), \quad \psi = (\delta_1 - \delta_2).$$

Appendix 2

THE SINE–GORDON EQUATION (SG)

The SG has the following form [A4]:

(a) in the laboratory frame of reference,

$$\varphi_{\tau\tau} - \phi_{\xi\xi} + \sin\varphi = 0 \tag{A.37}$$

(b) in the variables $\xi = x + t$, $\tau = t - x$,

$$\frac{\partial^2 \varphi}{\partial x \partial t} = \sin\varphi \tag{A.38}$$

Equation (A.38) is integrated with the help of the IST. The corresponding linear spectral problem for eigenvalues has the form

$$\begin{aligned} \psi_x^{(1)} &= \mathrm{i}\lambda\psi^{(1)} + \mathrm{i}(\varphi_x/2)\psi^{(2)} \\ \psi_x^{(2)} &= -\mathrm{i}\lambda\psi^{(2)} + \mathrm{i}(\varphi_x/2)\psi^{(2)}. \end{aligned} \tag{A.39}$$

The knowledge of the spectral data $a(\lambda)$, $b(\lambda)$ allows us to solve the problem of evolution of the initial condition in the SG. The corresponding solutions will be classified as follows.

(a) Let the coefficient $a(\lambda)$ be zero for $\lambda_1 = \mathrm{i}\lambda$. Then the solution is a soliton (antisoliton) of the Sg equation:

$$\varphi_s(x, t) = 4\tan^{-1}\exp[-\sigma(2\lambda(x - x_0) + t/2\lambda)], \tag{A.40}$$

where $\sigma = \pm 1$ stands for soliton or antisoliton.

In the laboratory frame of reference equation (A.41) has the form

$$\varphi_s(\xi, \tau) = 4\tan^{-1}\exp\{\sigma[\xi - \xi_0 - v\tau]/\surd(1 - v^2)\},$$
$$v = \frac{4\lambda^2 - 1}{4\lambda^2 + 1}, \qquad \xi_0 = (1 + v)x_0, \tag{A.41}$$

This describes a soliton moving with velocity v. For these solutions we derive a topological charge $Q = \sigma$,

$$Q = 1/2\pi \int_{-\infty}^{\infty} u_\xi \, d\xi. \tag{A.42}$$

(b) Let $a(\lambda)$ have a pair of zeros symmetric with respect to the imaginary axis

$$\lambda_1 = \mathrm{Re}\{\lambda\} + i\,\mathrm{Im}\{\lambda\}$$
$$\lambda_1 = -\mathrm{Re}\{\lambda\} + i\,\mathrm{Im}\{\lambda\}. \tag{A.43}$$

Then we have a solution describing a bound soliton and antisoliton (breather) state,

$$\varphi_B(x, t) = 4\tan^{-1}\left\{\frac{\mathrm{Im}\{\lambda\}}{|\mathrm{Re}\{\lambda\}|}\,\frac{\sin\left[\dfrac{\mathrm{Re}\{\lambda\}}{2|\lambda|^2}t - 2\,\mathrm{Re}\{\lambda x\} - \varphi_0\right]}{\cosh\left[2\,\mathrm{Im}\{\lambda(x - x_0) + \dfrac{\mathrm{Im}\{\lambda\}}{2|\lambda|^2}t\right]}\right\}. \tag{A.44}$$

In the laboratory frame of reference,

$$\varphi_B(\xi, t) = 4\tan^{-1}\left\{\frac{\mathrm{Im}\{\lambda\}}{\mathrm{Re}\{\lambda\}}\,\frac{\sin\left[\dfrac{\mathrm{Re}\{\lambda\}}{|\lambda|}\left(\dfrac{\tau - v\xi}{\surd(1 - v^2)} - \psi_0\right)\right]}{\cosh\left[\dfrac{\mathrm{Im}\{\lambda\}}{|\lambda|}\left(\dfrac{\xi - v\tau - \xi_0}{\surd(1 - v^2)}\right)\right]}\right\}. \tag{A.45}$$

The topological charge Q for (A.44 and 45) is zero.

Further we report the expressions for the energy of a soliton (or, as it usually called, a kink) and a breather. The energy is

$$E = \frac{1}{2}\int_{-\infty}^{\infty} dx(\varphi_t^2 + \varphi_x^2 + 4\sin^2(\varphi/2)). \tag{A.46}$$

For the soliton this is equal to

$$E_s = 8/\sqrt{(1 - v^2)}, \tag{A.47}$$

and for the breather this is written as follows:

$$E_b = 16(1 - v^2)^{-1/2} \sin \gamma, \qquad \gamma = \tan^{-1}[\mathrm{Im}\{\lambda\}/\mathrm{Re}\{\lambda\}].$$

Canonically conjugate variables have the forms

$$P_\lambda = (4/\pi\lambda)\ln(1/|a(\lambda)|^2), \qquad Q_\lambda = \arg b(\lambda).$$

Let us write out the expression for the SG Hamiltonian via the canonical variables:

$$H = \int_0^\infty (\lambda + 1/4\lambda)P_\lambda \, d\lambda + \sum_k 2(e^{-P_k} + 4e^{P_k})/\gamma$$

$$+ \sum_k (4/\gamma)\sin(\eta_k \gamma/16)(e^{-n_k/4} + 4e^{n_k/4}),$$

$$\eta_k = (16/\gamma)\arg \lambda_k; \quad n_k = 4\ln|\lambda_k|. \tag{A.48}$$

The first term in this expression is the contribution of the continuous spectrum to the energy, the second is the contribution of solitons, and the third term is that of breathers. Such a representation for the total energy is useful for calculating the energy of waves radiated by solitons and breathers in inhomogeneous and nonstationary media.

The Hamiltonian of a breather in terms of the variables γ, θ (where θ_τ is the breather self-frequency) is

$$H_b = \sin \gamma. \tag{A.49}$$

For $\gamma > \gamma_{sep} = \pi/2$ the breather separates into a free kink and antikink.

Next we will write out the equations of perturbation theory for solitons [A5]. A perturbed SGE is

$$\varphi_{tt} - \varphi_{xx} + \sin \varphi = \varepsilon R(\varphi).$$

In the adiabatic approximation, a one-soliton solution is sought in the form

$$\varphi_s(x, t) = 4\tan^{-1}\{\exp[x - \zeta(t)]/\sqrt{(1 - v^2)}\}. \tag{A.50}$$

To the first order in ε, the velocity v and coordinate of the soliton center $\zeta(t)$ have the following forms:

$$dv/dt = -(\varepsilon\sigma/4)(1 - v^2)^{1/2} \int_{-\infty}^{\infty} dz \, R(\varphi_s(z)) \operatorname{sech} z, \tag{A.51}$$

$$\mathrm{d}\zeta/\mathrm{d}t = v - (\varepsilon\sigma/4)v(1-v^2)\int_{-\infty}^{\infty} \mathrm{d}z\, zR(\varphi_s(z))\,\mathrm{sech}\, z,$$

$$z = (x-\zeta)/\sqrt{(1-v^2)}, \qquad \sigma = \pm 1. \tag{A.52}$$

The equations for the breather parameters are as follows [A6]:

$$\mathrm{d}\gamma/\mathrm{d}t = \varepsilon(1-v^2)^{1/2}(4\cos\gamma)^{-1}I_1, \tag{A.53}$$

$$\mathrm{d}v/\mathrm{d}t = -\varepsilon(1-v^2)^{3/2}(4\cos\gamma)^{-1}I_2, \tag{A.54}$$

$$\begin{aligned}\mathrm{d}\theta/\mathrm{d}t = {}& \cos\gamma(1-v^2)^{1/2} - \varepsilon(1-v^2)^{1/2}[v\cot\gamma I_3 + \cos^2\gamma(1-v^2)I_4] \\ & - I_5(4\sin\gamma\cos^2\gamma)^{-1},\end{aligned} \tag{A.55}$$

$$\mathrm{d}x_0/\mathrm{d}t = v + \varepsilon(1-v^2)(I_3 - v\tan\gamma\, I_4)(2\sin\gamma)^{-2}, \tag{A.56}$$

$$I_1 = \int_{-\infty}^{\infty} \cosh z \sin\Phi/(\cosh^2 z + A^2)R(\varphi_{\mathrm{B}}(z))\,\mathrm{d}z,$$

$$I_2 = \int_{-\infty}^{\infty} [\sinh z\cos\Phi/(\cosh^2 z + A^2)]R(\varphi_{\mathrm{B}}(z))\,\mathrm{d}z,$$

$$I_3 = \int_{-\infty}^{\infty} [z\cosh z\sin\Phi/(\cosh^2 z + A^2)]R(\varphi_B(z))\,\mathrm{d}z,$$

$$I_4 = \int_{-\infty}^{\infty} \cosh z\cos\Phi/(\cosh^2 z + A^2)R(\varphi_{\mathrm{B}}(z))\,\mathrm{d}z,$$

$$I_5 = \int_{-\infty}^{\infty} \cosh z\cos\Phi/(\cosh^2 z + A^2)R(\varphi_{\mathrm{B}}(z))\,\mathrm{d}z,$$

$$A = \tan\gamma\cos\varphi,$$

$$\Phi = (\mathrm{Re}\,\lambda/|\lambda|)(t - vx)/\sqrt{(1-v^2)} + \Phi_0.$$

The Jost functions for the one-soliton solution are ($\lambda_1 = \mathrm{i}v$)

$$\varphi = \frac{\mathrm{e}^{-ikz}}{\lambda + \mathrm{i}v}\begin{pmatrix} -\sigma v\,\mathrm{sech}\, z \\ \lambda - i\tanh z\end{pmatrix},$$

$$\psi = \frac{\mathrm{e}^{-ikz}}{\lambda + \mathrm{i}v}\begin{pmatrix} \lambda + \mathrm{i}v\,\mathrm{sech}\, z \\ \sigma v\,\mathrm{sech}\, z\end{pmatrix}. \tag{A.57}$$

Here $k = \frac{1}{2}(\lambda - (1/4\lambda))$.

The equation for Jost coefficients can be obtained as in Appendix 1. We have

$$\frac{\partial a}{\partial t}=\varepsilon\int_{-\infty}^{\infty}R[u(z)]\frac{\delta a}{\delta u_\tau(x)}\mathrm{d}x,$$

$$\frac{\partial b}{\partial t}=\mathrm{i}\omega(\lambda)b+\varepsilon\int_{-\infty}^{\infty}\mathrm{d}R[u(z)\frac{\delta b}{\delta u_\tau(x)}\mathrm{d}x. \qquad (A.58)$$

The variational derivatives $\delta a/\delta u_\tau, \delta b/\delta u_\tau$ can be found, for example, from the integral representation for the linear spectral problem in the variables τ, ξ, and are equal to

$$\delta a/\delta u_\tau=-\mathrm{i}(\psi_1\varphi_1-\psi_2\varphi_2)/4,\quad \delta b/\delta u_\tau=\mathrm{i}(\psi_1\varphi_2-\psi_2\varphi_2)/4. \qquad (A.59)$$

Substituting into (A.58), eqs. (A.59) and the expression for Jost functions (A.57), we obtain ($\tau\to t$)

$$\frac{\partial b}{\partial t}=\mathrm{i}\omega(\lambda)b-\frac{\mathrm{i}\varepsilon\gamma}{4(\lambda^2+v^2)}\exp(-\mathrm{i}\lambda\xi)\int_{-\infty}^{\infty}\mathrm{d}zR[u_s]$$

$$\cdot(\lambda^2-v^2-2\mathrm{i}\lambda v\tanh z)\exp(-\mathrm{i}k(\lambda)\gamma z). \qquad (A.60)$$

Using the solution for (A.60) we can calculate the contribution of a continuum spectrum to the energy, field momentum, etc. For example, the energy contribution can be written

$$E=\frac{4}{\pi}\int_{-\infty}^{\infty}\ln(1-|b|^2)\,\mathrm{d}x=\frac{4}{\pi}\int_{-\infty}^{\infty}|b(\lambda(k)|^2\,\mathrm{d}k,\quad k=\lambda-1/4\lambda. \qquad (A.61)$$

The kink–antikink interaction can be investigated by the PT, assuming $\varepsilon R=\sin(u_1+u_2)-\sin u_1-\sin u_2$ [A.8]. Then we obtain from (A.51) and (A.52) the equations for the distance between kinks $r(t)=\zeta_1-\zeta_2$, the average velocity $v=(v_1+v_2)/2$, and the relative velocity $p=v_1-v_2$

$$\mathrm{d}p/\mathrm{d}t=-8/\gamma^3\exp(-\gamma|r|),\quad \mathrm{d}r/\mathrm{d}t=p,\quad \mathrm{d}v/\mathrm{d}t=0,\quad \gamma=(1-v^2)^{-1/2} \qquad (A.62)$$

Here it is assumed $\gamma|r|\gg 1$, $|(v+p/4)p|\gamma^2\ll 1$, $|(v+p/4)p|\gamma^3 r\ll 1$. The Hamiltonian for (A.62) is

$$H=p^2/2-8\exp(-|r|). \qquad (A.63)$$

Perturbation theory can also be applied to a multidimensional case

[A.9]. Let us consider SG equation for a spherically symmetric case

$$u_{tt} - u_{rr} + \sin u = (D - 1)/r u_r - W'(u), \tag{A.64}$$

where D is the space dimension. When radius of a lump $\rho \gg 1$, the part of Laplacian $(D - 1)/r u_r$ can be treated as perturbation. The adiabatic has the form as in $(1 - D)$ case

$$u_s(r, t) = 4 \tan^{-1}\left\{\exp\left[-\frac{r - \rho(t)}{(1 - v^2)^{1/2}}\right]\right\}. \tag{A.65}$$

We can then apply the perturbation equations (A.51) and (A.52) to obtain

$$\begin{aligned}
\frac{dv}{dt} &= -\frac{D-1}{\rho}(1 - v^2) + a_\sigma(1 - v^2)^{3/2} + O(\varepsilon^2), \\
\frac{d\rho}{dt} &= v[1 + b_\sigma(1 - v^2)] + O(\varepsilon^2), \\
a_\sigma &= \frac{\sigma}{4}\int_{-\infty}^{\infty} W'[u_s] \operatorname{sech} z \, dz, \quad b_\sigma = \frac{\sigma}{4}\int_{-\infty}^{\infty} W'[u_s] z \operatorname{sech} z \tanh z.
\end{aligned} \tag{A.66}$$

This theory has been applied for the investigation $(2 - D)$ kink dynamics in LJJ with δ-like ring inhomogeneities [A.10] and for the calculation of emitted power by oscillating kink [A.11].

The perturbation theory can be also developed for the nearly integrable generalizations of NLSE and SGE.

Let us consider, for example, the generalized Maxwell–Bloch system [A.12].

$$\begin{aligned}
q_x &= \mathrm{i} a q_{\tau\tau} - \mathrm{i} g |q|^2 q = \langle \lambda \rangle + \varepsilon R, \\
\lambda_\tau &= \mathrm{i}\delta\lambda + qN, \\
N_\tau &= -1/2(q^*\lambda + \lambda^* q).
\end{aligned} \tag{A.67}$$

Here $\langle \cdots \rangle$ denotes the average over all frequency differences

$$\langle \lambda \rangle = \int \lambda(\delta) f(\delta) \, d\delta.$$

This system is completely integrable at $\varepsilon = 0$ and $a = g/2$. When $a = g = \varepsilon = \delta = 0$ this system is coincident with the SG system; when $\langle \lambda \rangle = \varepsilon = 0$ we have the NLS system. The soliton solution of an unperturbed system has the form

$$q = 4\eta \operatorname{sech} z \exp(\mathrm{i}\theta), \quad z = 2\eta(\tau - \tau_0(x)), \quad \theta = \mu z/\eta + \sigma(x),$$
$$\tau_0(x) = -(v/2\eta)x + \text{const}, \quad \sigma(x) = 2\mu\tau_0(x) + \delta(x) + \text{const},$$
$$\delta(x) = 4(\xi^2 - \eta^2)a - \omega_1, \quad v = 8\mu\eta a - \omega_2,$$
$$\omega_1 + \mathrm{i}\omega_2 = -\langle 1/(\delta + 2\lambda_1)\rangle, \quad \lambda_1 = \mu + \mathrm{i}\eta.$$

At $\varepsilon = 0$ we obtain, applying the above method, the following equation for the soliton parameters ($\mu(x), \eta(x), \tau_0(x)$ and $\sigma(x)$)

$$\begin{aligned}
\frac{\mathrm{d}\mu}{\mathrm{d}x} &= -\frac{1}{2}\int_{-\infty}^{\infty} \mathrm{d}z \frac{\tanh z}{\cosh z} \operatorname{Im}\{\varepsilon R \mathrm{e}^{\mathrm{i}\theta}\}, \\
\frac{\mathrm{d}\eta}{\mathrm{d}x} &= \frac{1}{2}\int_{-\infty}^{\infty} \frac{\mathrm{d}z}{\cosh z} \operatorname{Re}\{\varepsilon R \mathrm{e}^{\mathrm{i}\theta}\}, \\
\frac{\mathrm{d}\tau_0(x)}{\mathrm{d}x} &= -\frac{v}{2\eta} + \frac{1}{4\eta^2}\int_{-\infty}^{\infty} \mathrm{d}z \frac{z}{\cosh^2 z} \operatorname{Re}\{\varepsilon R \mathrm{e}^{\mathrm{i}\theta}\}, \\
\frac{\mathrm{d}\sigma}{\mathrm{d}x} &= 2\mu\tau_0(x) + \mu + \frac{1}{2\eta}\int_{-\infty}^{\infty} \frac{z}{\cosh z}(z \tanh z - 1) \operatorname{Im}\{\varepsilon R \mathrm{e}^{\mathrm{i}\theta}\}.
\end{aligned} \tag{A.68}$$

Appendix 3

KORTEWEG–DE VRIES EQUATION

The KdV equation has the form

$$u_t - 6uu_x + u_{xxx} = 0. \tag{A.69}$$

It is assumed that $u(x, t)$ is satisfied to the constraint

$$\int_{-\infty}^{\infty} |u(x, 0)|(1 + |x|)\,\mathrm{d}x < \infty. \tag{A.70}$$

The solution of the initial-value problem is given for initial conditions under (A.70) in according to IST [A.4], by the solution of the linear spectral problem

$$L\psi = -\,\mathrm{d}^2\psi/\mathrm{d}x^2 = u(x)\psi = \lambda\psi. \tag{A.71}$$

Here exists the discrete spectrum $\lambda_n < 0$, the eigenfunctions of which are square integrable and a double degenerating continuum spectrum $k^2 = \lambda > 0$. The values λ_n correspond to N-soliton states, the continuum to the radiation component of field.

The scattering problem for $u(x)$ can be given by the Jost functions φ, ψ, transition matrix $T(k)$

$$T(k) = \begin{Bmatrix} a(k) & b(k) \\ \bar{b}(k) & \bar{a}(k) \end{Bmatrix},$$

and the spectral parameter λ_n. The functions φ, ψ are defined by their asymptotic

$$\begin{aligned}\varphi &\to \exp(-ikx), \quad x \to \infty \\ \psi &\to \exp(-ikx), \quad x \to -\infty\end{aligned} \tag{A.72}$$

$$\varphi(x,k) = a(k)\psi(x,k) + b(k)\psi^*(x,k) \tag{A.73}$$

Wronskian is $W(\varphi,\varphi^*) = W(\psi,\psi^*) = 2ik$.

From (A.73) it follows that

$$a(k)^2 - b(k)^2 = 1 \tag{A.74}$$

The Jost coefficient $a(k)$ has simple zeros on the imaginary axis of a upper half λ plane. It can be shown that for N-soliton solution, $b = 0$ and the a coefficient can be represented thus:

$$a = \prod_{n=1}^{N} \frac{\lambda - \lambda_n}{\lambda - \lambda_n^*} \tag{A.75}$$

where $a(\lambda_n) = 0$, $b = b_n(\lambda_n)$ (and $\lambda_n = i\kappa_n$),

$$\varphi(i\kappa_n) = b_n \psi(x, -i\kappa_n). \tag{A.76}$$

For the Jost functions the following integral representation is valid

$$\psi(x,k) = \exp(-ikx) + \int_x^{\infty} K(x,y)\exp(-iky)\,dy. \tag{A.77}$$

The potential $u(x)$ is expressed by kernel $K(x,y)$ as

$$u(x) = -2dK(x,x)/dx. \tag{A.78}$$

The Gardner–Green–Kruskal–Miura formula for spectral data evolution takes the form:

$$\lambda_n(t) = \lambda_n(0), \quad a(\lambda,t) = a(\lambda,0), \quad b(\lambda,t) = b(\lambda,0)\exp(-8ik^3t).$$

$$b_n(\lambda,t) = b_n(0)\exp(-8i\kappa_n^3 t). \tag{A.79}$$

For $\lambda_1 = i\kappa$ the solution of GLM equation gives, for $u(x,t)$,

$$u(x,t) = -2\kappa^2 \operatorname{sech}^2[\kappa(x - 4\kappa^2 t - \varphi_0)], \tag{A.80}$$

where

$$\varphi = \frac{1}{2\kappa}\ln\frac{\beta}{2\kappa}; \quad \beta = \frac{b_1}{ia'(i\kappa)}, \quad a' = \left.\frac{\partial a}{\partial \lambda}\right|_{\lambda=\lambda_1}.$$

The polynomial integrals can be expressed by

$$2I_{j-1}[u] = \frac{2^{2(j+1)}}{2j+1}\sum_{l=1}^{N}\kappa_l^{2j+1} + 2^{2(j+1)}(-1)^{j+1}\frac{1}{\pi}\int_0^\infty k^{2j+1}\ln a\,\mathrm{d}k. \qquad (A.81)$$

For the Hamiltonian ($H = -I_1$) we obtain from (A.83)

$$H = -\frac{32}{5}\sum_{l=1}^{N}\kappa_l^5 + \frac{32}{\pi}\int_0^\infty k^4 \ln a(k)\,\mathrm{d}k. \qquad (A.82)$$

The variational derivatives for the Jost coefficients take the form

$$\frac{\delta a(k)}{\delta u(x)} = -\frac{1}{2ik}\psi^*(x,k)\varphi(x,k),$$

$$\frac{\delta b(k)}{\delta u(x)} = \frac{1}{2ik}\psi(x,k)\varphi(x,k). \qquad (A.83)$$

For a single soliton case (A.80) the Jost functions are equal to

$$\psi(x,k) = \frac{\mathrm{e}^{ikx}}{\kappa + ik}(ik + \kappa\tanh z)$$

$$\varphi(x,k) = a(k)\psi(x,k) = \frac{\mathrm{e}^{ikx}}{\kappa + ik}(ik - \kappa\tanh z). \qquad (A.84)$$

Then, using the approach described in Appendix 1 (A.32 and 33), we obtain for the Jost coefficients $a(k,t), b(k,t)$ the perturbation equations

$$\frac{\partial a}{\partial t} = \varepsilon\int_{-\infty}^{\infty}\mathrm{d}x\left\{\frac{\delta a}{\delta u(x)}R[u]\right\} = -\frac{\varepsilon}{2ik}\int_{-\infty}^{\infty}\mathrm{d}x\,R[u](\psi_s^*(x,k)\varphi_s(x,k))$$

$$\frac{\partial b}{\partial t} = 8ik^3 b + \varepsilon\int_{-\infty}^{\infty}\mathrm{d}x\left\{\frac{\delta b}{\delta u(x)}R[u]\right\} \qquad (A.85)$$

$$= 8ik^3 b + \frac{\varepsilon}{2ik}\int_{-\infty}^{\infty}\mathrm{d}x\,R[u](\psi_s(x,k)\varphi_s(x,k)).$$

Using the expression for the Jost functions as given by (A.57) we obtain

$$\frac{\partial a(k,t)}{\partial t} = \frac{i\varepsilon}{2\kappa k(k + i\kappa)^2}\int_{-\infty}^{\infty}(k^2 + \kappa^2\tanh^2 z)R[u_s]\,\mathrm{d}z, \qquad (A.86)$$

$$\frac{\partial b(k,t)}{\partial t} = 8\mathrm{i}k^3 b - \frac{\mathrm{i}\varepsilon\,\mathrm{e}^{-2\mathrm{i}k\xi}}{2\kappa k(k^2+\kappa^2)}\int_{-\infty}^{\infty} \mathrm{d}z\,\mathrm{e}^{-2\mathrm{i}kz/\kappa} R[u_s](k - \mathrm{i}\kappa\tanh z)^2. \tag{A.87}$$

The equation for a discrete spectrum data $\lambda_n(t)$, $b_n(t)$ is achieved by a slightly complicated route.

For $\lambda_n(t)$ we use the next formula

$$\delta a(\lambda_n) = \delta a|_{\lambda=\lambda_n} + \delta\lambda_n \left.\frac{\partial a}{\partial \lambda}\right|_{\lambda=\lambda_n} = 0. \tag{A.88}$$

Then

$$\left.\frac{\delta\lambda}{\delta u(x)}\right|_{\lambda n} = -\frac{\delta\lambda_n}{\delta u}\left.\frac{\partial a}{\partial\lambda}\right|_{\lambda n}. \tag{A.89}$$

Using (A.83) and (A.89) we obtain

$$\frac{\mathrm{d}\lambda}{\mathrm{d}t} = \varepsilon\int_{-\infty}^{\infty}\mathrm{d}x\,R[u]\frac{\delta\lambda_n}{\delta u} = -\varepsilon\int_{-\infty}^{\infty}\mathrm{d}x\,R[u]\frac{1}{a'|_{\lambda_n}}\left.\frac{\delta a}{\delta u}\right|_{\lambda n}$$

$$= \frac{\varepsilon}{2a'|_{\lambda_n}\kappa_n}\int_{-\infty}^{\infty}\mathrm{d}x\,R[u]\psi(x)\psi^*(x,k). \tag{A.90}$$

Substituting (A.84) we have for the soliton amplitude

$$\frac{\mathrm{d}\kappa}{\mathrm{d}t} = \frac{\varepsilon}{2k}\int_{-\infty}^{\infty}\mathrm{d}z\,R[u]\,\mathrm{sech}^2 z. \tag{A.91}$$

The equation for the soliton center ζ follows from the equation for b_1. Here we use the next trick. From (A.76) we have

$$\frac{\delta b_1}{\delta u} = \frac{\delta}{\delta u}\left(\frac{\varphi}{\psi}\right) = \frac{b_1}{2\kappa a'}\frac{\partial}{\partial k}[\varphi(\mathrm{i}\kappa)\psi(k) - \psi(\mathrm{i}\kappa)\varphi(k)]_{k=\mathrm{i}\kappa}.$$

Here we also use analog (A.88) for the ψ, φ functions.

In conclusion we have

$$\frac{\mathrm{d}b_1}{\mathrm{d}t} = 8\kappa^3 b_1 - \frac{\varepsilon b_1}{2\kappa^3}\int_{-\infty}^{\infty}\mathrm{d}z\,R[u_s(z)]\left(\tanh z + \frac{z+\kappa\xi}{\cosh^2 z}\right). \tag{A.92}$$

Substituting $b_1 = \varphi_1/\psi_1 = \mathrm{e}^{2\kappa\xi}$ in this equation we obtain the equation for soliton center $\xi(t)$

$$\frac{\mathrm{d}\xi}{\mathrm{d}t} = 4\kappa^2 - \frac{\varepsilon}{4\kappa^3}\int_{-\infty}^{\infty} \mathrm{d}z\, R[u_s(z)]\, \mathrm{sech}^2 z(z + \tfrac{1}{2}\sinh 2z). \tag{A.93}$$

Analogously, we can obtain the perturbation equations for the other nearly integrable systems. For example, for a perturbed KKS system the PT has been developed by Abdullaev *et al.* [A.14].

The perturbed KKS equation has the form:

$$u_{xt} + \tfrac{3}{2}(u_x)^2 u_{xx} + u_{xxxx} + h\sin u = \varepsilon R[u]. \tag{A.94}$$

Applying the method described above we obtain

$$\frac{\partial a}{\partial t} = \int_{-\infty}^{\infty} \tilde{\psi}^T \varepsilon \hat{R}[u]\varphi\, \mathrm{d}x, \tag{A.95}$$

$$\frac{\partial b}{\partial t} = 2b(4\mathrm{i}\lambda^3 + \mathrm{i}h/4\lambda) + \int_{-\infty}^{\infty} \hat{\psi}^T \varepsilon \hat{R}[u]\varphi\, \mathrm{d}x, \tag{A.96}$$

where

$$\varepsilon\hat{R}[u] = \begin{pmatrix} -\mathrm{i}\lambda_t & -\mathrm{i}\varepsilon R(u)/2 \\ \varepsilon R(u)/2 & \mathrm{i}\lambda_t \end{pmatrix}.$$

For the parameters of a single soliton solution

$$u = 4\tan^{-1}\{\exp[2\eta(x - \zeta_0)]\}, \quad \zeta_0 = vt - \delta_0.$$

$$\delta_0 = \ln\left(\frac{b(\mathrm{i}\eta)/a'(\mathrm{i}\eta)}{2\eta}\right), \quad v = 4\eta^2 + h/4\eta^2.$$

the system (A.94) and (A.95) takes the form

$$\frac{\mathrm{d}\eta}{\mathrm{d}t} = \frac{\varepsilon}{4}\int_{-\infty}^{\infty} R[u]\,\mathrm{sech}\, z\, dz, \tag{A.97}$$

$$\frac{\mathrm{d}\zeta_0}{\mathrm{d}t} = 4\eta^2 + \frac{h}{4\eta^2} + \frac{\varepsilon}{8\eta^2}\int_{-\infty}^{\infty} \mathrm{d}z\, z\,\mathrm{sech}\, z\, R[u], \quad z = 2\eta(x - \zeta_0). \tag{A.98}$$

REFERENCES

Introduction

I.1 Kosevich A. M., Ivanov B. A. and Kovalev A. S. *Nonlinear Magnetization Waves.* Kiev, Naukova Dumka, 1985 (In Russian).

I.2 Davydov A. S. Solutions in molecular systems. *Physica Scripta*, **20**, 387–394 (1979).

I.3 Zakharov V. E., Manakov S. V., Novikov S. P. and Pitaevsky L. P. *Theory of Solitons*, Consultants Bureau, N.Y. (1984).

I.4 Ablowitz M. J. and Segur H. *Solitons and Inverse Spectral Transform.* Philadelphia, SIAM (1980).

I.5 Lamb G. L., Jr. *Elements of Soliton Theory.* John Wiley, N.Y. (1980).

I.6 Dodd R. K., Eilbeck J. C., Gibbon J. D. and Morris H. C. *Solitons and Nonlinear Wave Equations.* Academic Press, London (1984).

I.7 Kivshar Yu. S. and Malomed B. A. *Rev. Mod. Phys.*, **61**, 763 (1990).

I.8 Newell A. C. *Solitons in Mathematics and Physics.* SIAM, Philadelphia (1985).

I.9 Hasegawa A. *Optical Solitons in Fibers.* Springer-Verlag, Heidelberg, 1990.

I.10 Abdullaev F. Kh., Darmanyan S. A. and Khabibullaev P. K. *Optical Solitons.* Springer-Verlag, 1993.

I.11 Abdullaev F. Kh. and Khabibullaev P. K. *Dynamics of Solitons in Inhomogeneous Condensed Media*, FAN, Tashkent, 1986.

I.12 Abdullaev F. Kh. *Phys. Rep.*, **179**, 1 (1989).

I.13 Gredeskul S. A. and Kivshar Yu. S. *Phys. Rep.*, **216**, 1 (1992).

Chapter 1

1.1 Zakharov V. E., Manakov S. V., Novikov S. P. and Pitaevsky L. P. *Theory of Solitons.* Consultant Bureau N.Y. (1984).

1.2 Ablowitz M. J. and Segur H. *Solitons and Inverse Scattering Transform.* Philadelphia, SIAM (1980).

1.3 Karpman V. I. *Nonlinear Waves in Dispersive Media.* Nauka, Moscow (1973).

1.4 Lamb G. L., Jr. *Elements of Soliton Theory.* John Wiley, N.Y. (1980).

1.5 Abramowitz M. and Stegun I. A. Handbook of Mathematical Functions. National Bureau of Standards, Washington D.C.

1.6 Zakharov V. E. and Shabat A. B. *Soviet Phys. JETP*, **34**, 62 (1972).

1.7 Lonngren K. and Scott A. P. (eds.). *Solitons in Action.* Academic Press, N.Y. (1978).

1.8 Whithem G. *Linear and Nonlinear Waves.* John Wiley, N.Y. (1974).

1.9 Faddeev L. D. and Korepin V. E. *Phys. Rep.*, **42**, 1–87 (1978).

1.10 Karpman V. I. *Phys. Lett.*, **A103**, 89 (1984).

1.11 Rubenstein J. *J. Math. Phys.*, **11**, 258 (1970).

1.12 Lakshmanan M. *Phys. Lett.*, **A61**, 53 (1977).

1.13 Takhtajan L. A. *Phys. Lett.*, **A64**, 235 (1977).

1.14 Kosevich A. M., Ivanov B. A. and Kovalev A. S. *Nonlinear Magnetization Waves.* Naukova Dumka, Kiev (1983) (in Russian).

1.15 Lakshmanan M., Ruijgrok T. W. and Thompson C. J. *Physica A*, 577 (1976).

1.16 Rajaraman R. *An Introduction to Solitons and Instantons in Quantum Field Theory.* North-Holland, Amsterdam (1982).

1.17 Krumhansl J. A. and Schrieffer J. R. *Phys. Rev.*, **B11**, 3565 (1975).

1.18 Makhankov V. G. *Comp. Phys. Comm.*, **21**, 1–49 (1980).

1.19 Bullough R. K. and Caudrey P. J. In Calogero F. (ed.). *Nonlinear Evolution Equations Solvable by the Spectral Transform.* Pitman, London (1978).

1.20 Makhankov V. G. *Phys. Rep.*, **35**, 1–128 (1978).

1.21 Kosevich A. M. Preprint IFM. Sverdlovsk (1976).

1.22 Davydov A. S. and Kislukha N. I. *Zh. Eks. Teor. Fiz.*, **71**, 1090 (1976) (In Russian).

1.23 Gorshkov K. A. and Ostrovsky L. A. *Physica*, **D3**, 428 (1981).

1.24 Kodama Y. and Ablowitz M. J. SIAM, **64**, 225 (1981).

1.25 Kaup D. J. and Newell A. C. *Proc. Roy. Soc. London*, **A361**, 413–446 (1978).

1.26 Karpman V. I. and Maslov E. M. *Sov. Phys. JETP*, **46**, 281 (1977).

1.27 McLaughlin D. W. and Scott. A. C. *Phys. Rev.*, **A18**, 1652 (1978).

1.28 Kivshar Yu. S. ILTP Preprint, Kharkov, No. 21–84 (1984).

1.29 Kivshar Yu. S. and Malomed B. A. *Rev. Mod. Phys.*, **61**, 763 (1990).

1.30 Keener J. P. and McLaughlin D. W. *Phys. Rev.*, **A10**, 777 (1977).

1.31 Kivshar Yu. S., Kosevich A. M. and Potemina. L. G. ILTP Preprint No. 37–85, 53pp., Kharkov (1985) (In Russian).

1.32 Maslov E. M. *Physica*, **D15**, 433 (1985).

1.33 Gorshkov K. A., Ostrovsky L. A. and Pelinovsky E. N. *Proc. IEEE*, **82**, 1511 (1974).

1.34 Zaslavsky G. M. *Stochasticity of Dynamical Systems.* Nauka, Moscow (1983) (In Russian).

1.35 Whitham G. B. *Linear and Nonlinear Waves*, Wiley, New York, 1974.

1.36 Abdullaev F. Kh. *Phys. Rep.*, **179**, (1989).

Chapter 2

2.1 Zabusky N. J. and Kruskal M. D. *Phys. Rev. Lett.*, **15**, 240–243 (1965).

2.2 Bishop A. R., Krumhansl J. A. and Trullinger S. F. *Physica*, **D1**, 1–44 (1981).

2.3 Abdullaev F. Kh., Abdumalikov A. A. and Umarov B. A. *Dokl. AN UzSSR* **3**, 25 (1989) (In Russian).

2.4 Olsen O. H. and Samuelsen M. R. *Phys. Rev. Lett.*, **48**, 1569 (1982).

2.5 Malomed B. A. *Physica*, **D27**, 113 (1987).

2.6 Lichtenberg. A. J. and Liberman M. A. *Regular and Stochastic Motion.* Springer, Heidelberg (1984).

2.7 Kivshar Yu. S. Preprint ILTP No. 21–84, Kharkov (1984).

2.8 Chen H. H. and Liu C. S. *Phys. Rev. Lett.*, **37**, 693 (1976).

2.9 Karpman V. I. and Maslov E. M. *Phys. Fluids*, **25(9)**, 1686 (1982).

2.10 Kosevich A. M. Preprint ILTP, Kharkov (1989). (*Physica* **D** (1990), in press).

2.11 Abdullaev F. Kh. *Lebedev Inst. Reports.* (Consultants Bureau, New York) Vol. 1, p. 52 (1983).

2.12 Gorshkov K. A., Ostrovsky L. A. and Papko V. V. Solitons in radiophysics. *Physica Scripta*, **20**, 357 (1979).

2.13 Fabrikant M. V. *Wave Motion*, **2**, 355 (1980).

2.14 Kodama Yu. and Hasegawa A. *Opt. Lett.* **7**, 285 (1982).

2.15 Nayfeh A. H. *Perturbation Methods.* Wiley, N.Y. (1982).

2.16 Gorshkov K. A. Doctoral Dissertation, Gorky (1981).

2.17 Djumaev M. R. Doctoral Dissertation, Tashkent (1988).

2.18 Kolomensky A. A. and Lebedev A. N. *Theory of Cyclic Accelerators.* Fizmatgiz, Moscow (1962).

2.19 Hasegawa A. and Kodama Yu. Preprint Math. Dep. OSU, Ohio, USA (1990).

2.20 Abdullaev F. Kh. and Abdumalikov A. A. *Sov. Phys.-Dokl.*, **31**, 502 (1986).

2.21 Bogdan M. M., Kosevich A. M. and Manzhos I. V. *Low Temperature Physics*, **11**, 991 (1985).

2.22 Karpman I. V., Maslov E. M. and Solov'ev V. V. *Sov. Phys.-JETP*, **57**, 167 (1983).

2.23 Kivshar Yu. S. and Malomed B. A. *Rev. Mod. Phys.*, **61**, 763 (1989).

2.24 Abdullaev F. Kh., Abrarov R. M. and Darmanyan S. A. *Opt. Lett.*, **14**, 48 (1989).

2.25 Karpman V. I. and Solov'ev V. V. Preprint IZMIRAN n34(262) Moscow (1979); *Physica* **D3**, 142–164 (1981).

2.26 Abdullaev F. Kh., Darmanyan S. A. and Khabibullaev P. K. *Optical Solitons.* FAN, Tashkent (1987) (In Russian). (Translated by Springer-Verlag, 1993).

2.27 Wright E. M., Stegeman G. I. and Wabnitz S. *Phys. Rev.*, **A40**, 4455 (1989).

2.28 Darmanyan S. A. *Opt. Comm.*, **90**, 301 (1992).

2.29 Chirikov B. V. *Phys. Rep.*, **52**, 263 (1979).

2.30 Ueda T. and Kath W. L. *Phys. Rev.*, **A42**, 563 (1990).

2.31 Herrera J. J. *J. Phys. A: Math. Gen.*, **17**, 95 (1984).

2.32 Abdullaev, F. Kh. and Nyazov B. A. *Sov. J. Techn. Phys.*, **52**, 2631 (1982) (In Russian).

2.33 Pagano S., Salerno M. and Samuelsen M. R. *Physica*, **26D**, 396 (1987).

2.34 Abdullaev F. Kh., Abdumalikov A. A. and Stamenkovich L. *Phys. St. Solidi* (b), **120**, 33 (1984).

2.35 Rodrigez-Plaza M. J. and Vazquez L. *Phys. Rev.*, **B41**(16), 11437 (1990).

2.36 Newell A. C. *Solitons in Mathematics and Physics.* University of Arizona, Soc. Ind. and Appl. Math. (1985).

2.37 Grimshaw R. J. *Fluid Mech*, **42**, 639 (1970).

2.38 Pelinovsky E. N. *PMTF*, **2**, 68 (1971) (In Russian).

2.39 Ko K. and Kuehl H. H. *Phys. Rev. Lett*, **40**, 233 (1978).

2.40 Scharf R. and Bishop A. R. *Phys. Rev.*, **A46**, 2973 (1992).

2.41 Wabnitz S. *Electr. Lett.*, **29**, 1711 (1993).

2.42 Akhmediev N. N. and Ankiewicz A. *Phys. Rev. Lett.*, **70**, 2395 (1993).

Chapter 3

3.1 Feinberg E. L. In *Problems of Modern Physics.* Nauka Publ. Moscow 1978 (In Russian).

3.2 Kivshar Yu. S., Kosevich A. M. and Chubykalo D. A. *Phys. Lett.*, **A125**, 35 (1987).

3.3 Acevec A. B., Newell A. C. and Moloney J. A. *Phys. Rev.*, **A39**, 1809 (1989); **A39**, 1828 (1989).

3.4 Mkrtchyan G. S. and Schmidt V. V. *Sol. St. Commun.*, **30**, 791 (1979).

3.5 Kivshar Yu. S. and Malomed B. A. *Rev. Mod. Phys.*, **61**, 763 (1989).

3.6 Abdullaev F. Kh. and Djangiryan R. G. *Izv. Vuz'ov: Radiofizika*, **53**, 2307 (1982) (In Russian).

3.7 Umarov B. A. Doctoral Dissertation, Tashkent, 1986.

3.8 Ginzburg V. L. *Problems of Physics and Astrophysics.* Nauka, Moscow (1984) (In Russian).

3.9 Bolotovsky B. M. *Uspeki Fiz. Nauk*, **136**, 501 (1982) (In Russian) (*Sov. Phys.-Uspekhi*).

3.10 Abdullaev F. Kh. and Djañgiryan R. G. *Izv. Vuz'ov: Radiofizika*, **27**, 985 (1984) (In Russian).

3.11 Aranson I. S. Preprint Inst. Appl. Phys. N240, Gorky, USSR (1989).

3.12 Kow P. K., Tzinzadze N. L. and Tzhakaya D. D. *JETP*, **82**, 1449 (1982) (In Russian).

3.13 McLaughlin D. W. and Scott A. C. *Phys. Rev.*, **A18**, 1652 (1978).

3.14 Baryakhtar V. G., Ivanov B. A. and Sukstansky A. L. *JETP*, **75**, 2183 (1978) (In Russian).

3.15 Scott J. C., Zimm C. B. and Sarhanji A. *J. Appl. Phys.*, **53**(3), 2768 (1982).

3.16 Abdullaev F. Kh. and Djangyryan R. G. *Phys. St. Solidi* (b), **129**, K45 (1985).

3.17 Abdullaev F. Kh. and Abdumalikov A. A. *Izv. AN UzSSR*, 31 (1988) (In Russian).

3.18 Kosevich A. M. *Theory of Nonlinear Lattices.* Sverdlovsk, 1978. (In Russian).

3.19 Tappert F. D. and Zabusky N. *J. Phys. Rev. Lett.*, **27**, 1774 (1971).

3.20 Pelinovsky E. N. *PMTF*, **N6**, 80 (1971) (In Russian).

3.21 Chukbar K. V. and Yankov V. V. *Plasma Physics*, **3**, 1398 (1977) (In Russian).

3.22 Tornhill S. and Ter-Haar D. *Phys. Rep.* **43**, 45 (1978).

3.23 Kurin V. V. and Friman G. M. *Fiz. Plasmy.* **7**, 716 (1981) (In Russian).

3.24 Djumaev M. R. PhD Dissertation, Tashkent, 1988.

3.25 Abdullaev F. *et al.* In *Optical Solitons*, World Scientific Singapore, 8 (1991).

3.26 Abdullaev F. Kh. and Djangiryan R. G. *JTP*, **26**, 281 (1981). (In Russian).

3.27 Abdullaev F. Kh. and Umarov B. A. *Phys. Lett.*, **A160**, 429 (1991).

3.28 Abdullaev F. Kh., Abdumalikov A. A. and Umarov B. A. *Phys. Lett.*, **A171**, 125 (1992).

Chapter 4

4.1 Sinai Ya. G. In *Nonlinear Waves* p. 192, Nauka, Moscow, 1979 (In Russian).

4.2 Rytov S. M., Kravtzov Yu. A. and Tatarsky V. I. *Introduction to Statistical Radiophysics*, Vol. 2, p. 463. Nauka, Moscow (1978) (In Russian).

4.3 Howe M. S. *J. Fluid Mech.*, **45**, 785–804 (1971).

4.4 Gurbatov A. N., Pelinovsky E. N. and Saichev A. I. *Izv. Vuz'ov: Radiofizika*, **21**, 1485 (1978) (In Russian).

4.5 Abdullaev F. Kh. *Izv. Vuz'ov: Radiofizika*, **25**, 756–758 (1982) (In Russian).

4.6 Pelinovsky E. N. In *Nonlinear Waves.* Nauka, Moscow (1979) (In Russian).

4.7 Abdullaev F. Kh. Teor. Math. Phys., **51**, 454–459 (1982) (In Russian).

4.8 Klyatzkin V. I. *Stochastic Equations and Waves in Randomly Inhomogeneous Media*, Nauka, Moscow (1980) (In Russian).

4.9 Benilov E. S. and Pelinovsky E. N. *Doklady USSR Academy of Sciences*, **301**, 1100–1103 (1988) (In Russian) (*Sov. Phys.-Doklady*).

4.10 Wadati M. *J. Phys. Soc. Japan*, **52**, 2642–2648 (1983).

4.11 Wadati M. and Akutsu Y. *J. Phys. Soc. Japan*, **53**, 3342–3350 (1984).

4.12 Beziers I. In *Nonlinear Electromagnetics*. Academic Press, N.Y. (1980).

4.13 Khikmatov N. A. Doctoral Dissertation, Tashkent (1987).

4.14 Maimistov A. I. and Manykin E. A. *Izv. Vuz'ov: Fizika*, **4**, 91–97 (1987).

4.15 Abdullaev F. Kh. *Lebedev Institute Reports on Physics*, **10**, 3–8 (1982).

4.16 Abdullaev F. Kh. *et al. Contributions of III International Symposium on Some Problems of Statistical Mechanics*, Dubna, JINR 3–7 (1985).

4.17 Abdullaev F. Kh. Darmanyan S. A. and Djumaev M. *Izv. UzSSR AS*; *ser. fiz. mat. nauk*, **6**, 34–39 (1986); *Phys. Lett.* **A108**, 51 (1989).

4.18 Dianov E. M. *Doklady AN USSR*, **283**, 1324–1346 (1985).

4.19 Meerson B. I. In *Wave and Diffractions*, pp. 126–128, Moscow (1981) (In Russian).

4.20 Filonenko N. N. *Doklady AN USSR*, **189**, 1208–1211 (1970) (In Russian).

4.21 Abdullaev F. Kh. *Fiz. Met. Metallov.* **57**, 450–456 (1984) (In Russian).

4.22 Bass F. G., Konotop V. V. and Sinitsyn Yu. A. *JETP*, **88**, 541–549 (1985) (In Russian).

4.23 Abdullaev F. Kh., Abdumalikov A. A. and Umarov B. A. *Izv Vuz'ov: Fizika*, **4**, 57–62 (1989) (In Russian).

4.24 Aubry J. J. *Chem. Phys.* **62**, 3217 (1975).

4.25 Krumhansl J. A. and Schrieffer J. R. *Phys. Rev.*, **B11**, 3535–3545 (1975).

4.26 Landau L. D. and Lifshitz E. M. *Quantum Mechanics*, p. 546, Moscow (1963).

4.27 Umarov B. A. Doctoral Dissertation, Tashkent, 1986.

4.28 Umarov B. A. and Khikmatov N. A. *Izv UzSSR AS ov: Fiz. mat. nauk*, **1**, 62–64 (1989) (In Russian).

4.29 Kivshar, Yu. S. Preprint KhFTINT, 21–85, 60 (1984).

4.30 Kivshar Yu. S. and Konotop V. V. *JETP*, **59**, 1–9 (1989).

4.31 Bass F. G., Vatova L. B. and Konotop V. V. Preprint IRE, 364, 36, Kharkov (1989).

4.32 Abdullaev, F. Kh. and Khabibullaev P. K. *Dynamics of solitons in inhomogeneous condensed media.* FAN Publ., 200, Tashkent (1986) (In Russian).

4.33 Abdullaev F. Kh. Abdumalikov A. A. and Stamenkovich L. J. *Phys. Stat. Solid* (b), 120, 33–37 (1983).

4.34 Collins M. A. *et al. J. Phys. Rev*, **B19**, 3630–3644 (1979).

4.35 Stratonovich R. L. *Selected Topics in Noise.* Sovetskoe Radio Publ., 558, Moscow (1961).

4.36 Abdullaev F. Kh. and Khikmatov N. A. *Uzv. VUZ'ov: Radiofizika*, **24**, 408–412 (1983) (In Russian).

4.37 Bass F. G. *et al. Z. Phys.* **B65**, 209–223 (1986).

4.38 Abdullaev F. Kh. and Khikmatov N. A. *JETP*, 56, 1206–1207 (1986) (In Russian).

4.39 Zaslavsky G. M., *JETP*, **66**, 1632–1639 (1974) (In Russian).

4.40 Abdullaev F. Kh. and Abdumalikov A. A. *J. Phys. Stat. Solidi* (b) **114**, 650–655 (1982).

4.41 Makhankov V. G. *Elementary Particles and Atomic Nuclei* **14**, 123–180 (1981) (In Russian).

4.42 Eilbeck J. C. and McGuire G. J. *J. Comp. Phys.*, **19**, 43–57 (1975).

4.43 Bykov V. V. *Digital Modelling in Statistical Radiophysics.* Sovetskoe Radio Publ., 326, Moscow (1971) (In Russian).

4.44 Zhao Y. *Optics Commun*, **68**, 21–24 (1988).

4.45 Akhmanov S. A. Vysloukh V. A. and Chirkin A. S. *Uspekhi Fiz. Nauk* **149**, 449–510 (1986) (Sov. Phys.-Uspekhi).

4.46 Malomed B. A. *Phys. Scripta* **38**, 66 (1988).

4.47 Malomed B. A. Doctoral Dissertation, Kiev (1989).

4.48 Caputo J. G., Newell A. C. and H. Shelley. In *Integrable Systems and Applications.* Springer-Verlag, Heidelberg (1990).

4.49 Gredeskul *et al. Phys. Rev.*, **A45**, 8867 (1992).

4.50 Mineev M. B., Feigelman M. V. and Schmidt V. V. *Sov. Phys. JETP*, 54, 155 (1981).

4.51 Bass F. *et al. Phys. Rep.*, **157**, 63 (1988).

4.52 Abdullaev F. Kh. *Dynamical Chaos of Solitons.* FAN Tashkent (1990) (In Russian).

Chapter 5

5.1 Abdullaev F. Kh. *Lebedev Inst. Rep.* **1**, 52, Moscow (1983) (In Russian).

5.2 Chirikov B. V. *Phys. Rep.* **52**, 263 (1979).

5.3 Melnikov V. K. *Mosc. Math. Soc. Trans.* **12**, 3 (1963) (In Russian).

5.4 Aranson I. S., Gorshkov K. A. and Rabinovich M. I. Preprint IPFAN SSSR, N51, Gorky (1982) (In Russian).

5.5 Zaslavsky G. M. *Stochasticity of Dynamical Systems.* Nauka, Moscow (1985).

5.6 Aranson I. S., Gorshkov K. A. and Rabinovich M. I. *JETP*, **84**, 929 (1984).

5.7 Davydov A. S. *Theory of Molecular solitons.* Kiev (1985).

5.8 Abdullaev F. Kh. and Khykmatov N. A. *JTP*, **56**, 987 (1985) (In Russian).

5.9 Karpman V. I. *Phys. Lett.* **88A**, 207 (1982).

5.10 Nozaki K. *Phys. Rev. Lett.*, **49**, 1883 (1982).

5.11 Malomed B. A. *Physica*, **27D**, 113 (1987).

5.12 Abdullaev F. Kh., Darmanyan S. A. and Umarov B. A. *Phys. Lett.*, **108A**, 53 (1985).

5.13 Karpman V. I., Maslov E. M. and Solov'ev V. V. *JETP*, **84**, 288 (1983).

5.14 Mikeska H. J. *J. Phys. C: Solid St. Phys.*, **11**, 129 (1978).

5.15 Abdullaev F. Kh. *Phys. Rep.*, **179**, 1 (1989).

5.16 Abdullaev F. Kh., Darmanyan S. A. and Umarov B. A. *Phys. Rev.*, **A41**, 4498 (1990)

5.17 Abdullaev F. Kh. and Abdumalikov A. A. In *Nonlinearity with Disorder.* Springer-Verlag, Heidelberg (1990).

5.18 Akhmediev N. N., Eleonsky V. M. and Kulagin N. N. *Sov. Phys. JETP*, **62**, 894 (1985).

5.19 Bishop A. R., Flesh R., Forest M. G., Mclaughlin D. W. and Overman E. D. *SIAM J. Math. Anal.*, **21**, 1511 (1990).

5.20 Ablowitz M. J. and Herbst B. M. *SIAM J. Appl. Math.*, **50**, 339 (1990).

5.21 McLaughlin D. W. In *Important Development in Soliton Theory.* A. Fokas, V. Zakharov Eds. Springer-Verlag, Heidelberg (1993).

5.22 Abdullaev F. Kh. *Dynamical Chaos of Solitons.* FAN Tashkent (1990). (In Russian).

5.23 Ablowitz M. J. and Clarkson P. A. *Solitons, Nonlinear Evolution Equations and Inverse Scattering.* Cambridge University Press (1991).

Appendices

A.1 Karpman V. I. and Maslov E. M. *Sov. Phys.-JETP*, **46**, 281 (1977).

A.2 Kaup D. J. and Newell A. C. *Proc. Roy. Soc.*, **A361**, 413 (1978).

A.3 Malomed B. A. *Physica*, **D27**, 113 (1987).

A.4 Zakharov V. E. *et al. Theory of Solitons. The Inverse Scattering Method.* Nauka, Moscow (English translation, Consultants Bureau, N.Y.)

A.5 McLaughlin D. W. and Scott A. C. In Longren K. and Scott A. C. (eds.), *Solitons in Action.* Academic Press, N.Y. (1978).

A.6 Karpman V. I., Maslov E. M. and Solov'ev V. V. *Sov. Phys.-JETP*, **57**, 167 (1983).

A.7 Karpman V. I. and Solov'ev V. V. *Physica*, **D3**, 142 (1981).

A.8 Karpman V. I. *Phys. Lett.*, **88A**, 207 (1982).

A.9 Maslov E. *Physica*, **D15**, 433 (1985).

A.10 Abdullaev F. Kh. and Umarov B. A. In *Nonlinearity with Disorder*, Springer-Verlag, Heidelberg (1992).

A.11 Malomed B. A. *Physica D*, in press.

A.12 Abdullaev F. Kh. and Tadjimuratov S. A. *Sov. Phys. Doklady*, **36**, 57 (1991).

A.13 Faddeev L. D. and Takhtadjan L. A. *Hamiltonian Methods in the Theory of Solitons.* Springer-Verlag, Heidelberg (1987).

A.14. Abdullaev F. Kh., Abdumalikov A. A. and Tsoi E. N. *Phys. Lett.*, **A151**, 221 (1990).

INDEX